Motor Control, Learning and Development

BIOS INSTANT NOTES

Series Editor: B. D. Hames, School of Biochemistry and Molecular Biology, University of Leeds, Leeds, UK

Biology

Animal Biology, Second Edition	9781859963258
Biochemistry, Third Edition	9780415367783
Bioinformatics	9781859962725
Chemistry for Biologists, Second Edition	9781859963555
Developmental Biology	9781859961537
Ecology, Second Edition	9781859962572
Genetics, Third Edition	9780415376198
Human Physiology	9780415355469
Immunology, Second Edition	9781859960394
Mathematics & Statistics for Life Scientists	9781859962923
Medical Microbiology	9781859962541
Microbiology, Third Edition	9780415390880
Molecular Biology, Third Edition	9780415351676
Motor Control and Development	9780415391399
Neuroscience, Second Edition	9780415351881
Plant Biology, Second Edition	9780415356435
Sport & Exercise Biomechanics	9781859962848
Sport & Exercise Physiology	9781859962497

Chemistry

Analytical Chemistry	9781859961896
Inorganic Chemistry, Second Edition	9781859962893
Medicinal Chemistry	9781859962077
Organic Chemistry, Second Edition	9781859962640
Physical Chemistry	9781859961940

Psychology

Sub-series Editor: Hugh Wagner, Dept of Psychology, University of Central Lancashire, Preston, UK

Cognitive Psychology	9781859962237
Physiological Psychology	9781859962039
Psychology	9781859960974
Sport & Exercise Psychology	9781859962947

Motor Control, Learning and Development

Andrea Utley
Director, Centre for Sport and Exercise Sciences,
University of Leeds, Leeds, UK

Sarah Astill
Roberts Fellow (Life Science Interface),
School of Health Professions and Rehabilitation Sciences,
University of Southampton, Southampton, UK

Published by:
Taylor & Francis Group
In US: 270 Madison Avenue,
New York, NY 10016
In UK: 2 Park Square, Milton Park
Abingdon, Oxon OX14 4RN

First published 2008

ISBN: 978-0-4153-9139-9

This book contains information obtained from authentic and highly regarded sources. Reprinted material is quoted with permission, and sources are indicated. A wide variety of references are listed. Reasonable efforts have been made to publish reliable data and information, but the author and the publisher cannot assume responsibility for the validity of all materials or for the consequences of their use.

A catalog record for this book is available from the British Library.

Library of Congress Cataloging-in-Publication Data

Utley, Andrea.
Motor control, learning and development / Andrea Utley, Sarah Astill.
p. cm. -- (Bios instant notes)
Includes bibliographical references and index.
ISBN 978-0-415-39139-9
1. Motor ability. I. Astill, Sarah. II. Title.

BF295.U85 2008
152.3'34--dc22

2007052427

Editor: Elizabeth Owen
Editorial Assistant: Sarah Holland
Senior Production Editor: Simon Hill
Typeset by: Phoenix Photosetting, Chatham, Kent
Printed by: Cromwell Press Ltd

Printed on acid-free paper

10 9 8 7 6 5 4 3 2 1

Taylor & Francis Group
is the Academic Division of T&F Informa plc.

Visit our web site at http://www.garlandscience.com

CONTENTS

ABBREVIATIONS

ABC	assessment battery for children
ACh	acetylcholine
AE	absolute error
AMC	age-matched control
CE	constant error
CI	contextual interference
CNS	central nervous system
CRT	choice reaction time
DCD	developmental coordination disorder
DRT	discrimination reaction time
EEG	electroencephalography
EMG	electromyography
fMRI	functional magnetic resonance imaging
FRT	fractionated reaction time
GMP	generalized motor program
ID	index of difficulty
KP	knowledge of performance
KR	knowledge of results
LED	light-emitting diode
LOP	levels-of-processing (model)
LTM	long-term memory
MRI	magnetic resonance imaging
MT	movement time
NGST	neuronal group selection theory
PET	positron emission tomography
PNS	peripheral nervous system
PRP	psychological refractory period
RMS	root mean square
RRT	recognition reaction time
RT	reaction time
SOA	stimulus onset asynchrony
S–R	stimulus–response
SRT	simple reaction time
STM	short-term memory
STSS	short-term sensory storage
SUSOPs	sustained operations
TKA	total knee arthroplasty
TPA	time to peak acceleration
TPV	time to peak velocity
TT	total time
VE	variable error

Preface

In writing this book our main aim was to produce a resource for undergraduates undertaking one of the many sports science or sport and exercise science degrees that are now available. We also aimed to produce a book that would be useful to individuals taking sports coaching courses or training to be physical education teachers or physiotherapists. The early sections set the scene by defining the field and discussing measurement. This leads to theories of control and sections that deal with how we control movement and how we learn to control movement. The final section considers the development of motor skills, as we felt that this was an important aspect that needed to be considered and that is rarely covered in texts on motor control. The topics covered in this text provide foundation knowledge vital for any individual who is working in the movement context as a teacher, coach or therapist. Examples are given from recent and relevant research, further reading is suggested, and activities provided to strengthen understanding. We have also attempted to provide examples from a range of sports and activities that are especially relevant to the UK and Europe. Each section can be read in isolation, but links are made and related topics highlighted. We hope you enjoy this book and that the content helps improve your practice and knowledge in relation to sport, education and rehabilitation.

Andrea Utley

Acknowledgments

We would like to acknowledge the support of Rachel Hartley for proofreading the original texts and Helen Turton for taking a number of the photographs that appear in the book. In addition, we would like to thank Laura and Nigel for their support and patience.

Dedications

To my sister Beverley
Andrea Utley

To my parents
Sarah Astill

Section A

WHAT IS MOTOR CONTROL?

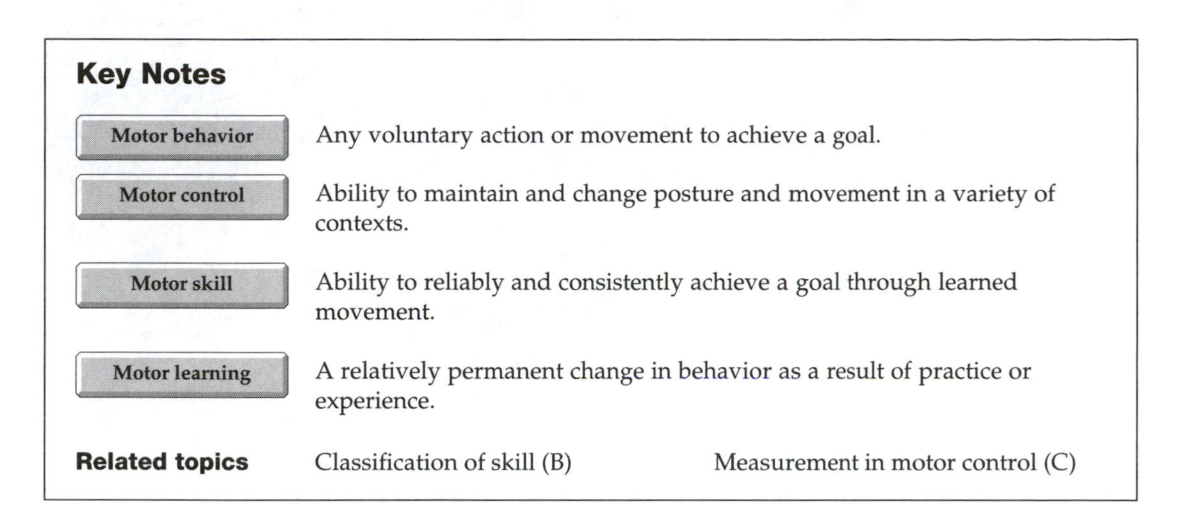

Key Notes

Motor behavior	Any voluntary action or movement to achieve a goal.
Motor control	Ability to maintain and change posture and movement in a variety of contexts.
Motor skill	Ability to reliably and consistently achieve a goal through learned movement.
Motor learning	A relatively permanent change in behavior as a result of practice or experience.

Related topics Classification of skill (B) Measurement in motor control (C)

In our daily lives we produce a whole host of movements that are crucial to our independence, interactions with the world and our personal safety. Motor control is the study of movements, the mechanisms that enable movements to be produced and the processes that underlie control, skill acquisition and retention. When studying motor control we are asking questions about what needs to be controlled, how we learn to do this, and how we are able to coordinate the vast range of simple and complex movements that we perform. The term motor control also reflects a multidisciplinary approach to asking and answering questions about control. Knowledge of motor control and motor learning should form a fundamental part of any individual's theoretical background who is involved in sport, physical activity and rehabilitation. This section provides an overview of the field of motor control, defines the terms used and considers why we need to study motor control, learning and development.

Definition of terms

It is through movement that we interact with the world either by moving around in different contexts or handling objects, or by dealing with other people (*Fig. 1*). In order to do this we have to control a mass of different movements that vary in complexity and speed while dealing with multiple or related inputs. A range of terms is used to define movement, and as a starting point to the remainder of this book it is important that these are clearly defined. We have used the term motor control as a starting point; however the term motor behavior is often used (see Section B for further definitions). Many researchers use the term motor behavior to describe any motor action or movement that is used to complete a task or achieve a goal. The study of motor behavior is then divided into three areas: motor control, motor learning and motor development. It is useful to understand the distinction between these areas.

Motor control

The starting point in defining motor control is to remember that it is the study of the nature and cause of movements or actions. When studying motor control we

(a)

(b)

Fig. 1. In our daily lives we produce a whole host of movements such as writing, sitting, standing (a) and more complex movements such as whitewater kayaking (b).

are studying postures and movements and the mechanisms that enable us to move. There are three key aspects to motor control that have to be taken into account. Firstly, motor control is concerned with the study of action. When we examine motor control we often consider a particular action such as running, catching, or picking up an object. However, it must be remembered that an important part of movement is the interaction between the environment and the individual. As we move, we use sensory information about our position in the environment and the position of our body in relation to the environment and other body parts. Motor control is therefore also concerned with the study of perception. Finally we must also remember that movement involves cognitive processes in order to organize perception and action. Therefore motor control is also concerned with the study of cognition. Another feature of the study of motor control is the formation of theories, the testing of theories, and the use of models to understand movement. A theory of motor control is a group of abstract ideas about the nature and the cause of movement. A model is a representation of something, usually a simplified version of the real thing. The better the model, the better it will predict how the real thing will behave in a real situation. Throughout this book we will consider theories and models of control in order to critically assess human motor control.

Motor learning

Motor learning is an area that has received a mass of attention from a range of researchers who are interested in examining how we learn and retain movement skills. Learning is defined as a change in a person's capability to perform a skill. Wolpert *et al.* (2001) state that learning involves a change in behavior that occurs as a result of interaction with the environment that is distinct from maturation. It must be remembered that this is inferred from a relatively permanent improvement in performance as a result of practice or experience. Two terms that should not be confused are learning and performance (see Section B). Performance is observable; you can see a footballer take a penalty or watch a climber; learning, however, cannot be directly observed and can only be inferred from the nature of the movement produced (*Fig. 2*).

Fig. 2. Performance is observable; you can watch a climber complete a route for example.

Motor development

Motor development refers to the process of change that a person passes through as they grow and mature. Development, according to Keogh and Sugden (1985, p.6) is an 'adaptive change towards competence'. The human adult is capable of performing a variety of motor skills and demonstrates a high level of control and dexterity. The development of motor control is complex and involves the development of many behaviors (see Section J). Motor development in children is so rapid and so apparent that we sometimes forget that motor development is a lifelong process – learning to walk for example. During the development of motor control, children are establishing the movement patterns and control synergies that will be used throughout life. This development has implications when studying how movements are coordinated and how this is influenced by task constraints and the environmental context.

Origins of the field

The field of motor control emerged from two separate fields of study; neurophysiology and psychology. These fields developed separately and it was only in the 1970s that they came together and motor control became an area of study. Many of the early studies examined tasks such as hand movements (Bowditch and Southard, 1882) and learning tasks such as typing and Morse code (Bryan and Harter, 1899; Hill *et al.*, 1913). Bryan and Harter's (1899) first formal study of skill acquisition looked at the way people learn to communicate with Morse code. They noted that:

- learning is discontinuous in nature (plateaus)
- we tend to learn small units and combine these into larger units, then even larger units (chunking).

The second point is still largely credible today. It was Woodworth (1899), however, who was the first to attempt to fully understand motor skills by comparing fast and slow movements. He asked subjects to move a pencil back and forth through a slit, with the direction reversing once two marks had been passed. The rate of moving the pencil was controlled by a metronome and subjects performed one set with their eyes open and another set with their eyes closed. He discovered that when subjects had their eyes closed, the mean absolute error remained more or less constant as velocity decreased. When subjects had their eyes open the mean absolute error decreased as velocity decreased. Therefore accuracy improved as movements slowed in the eyes-open condition but not in the eyes-closed condition. Woodworth concluded that in the eyes-closed condition subjects' movements were entirely preprogramed; he referred to this as an initial impulse. In the eyes-open condition he concluded that the movements were both preprogramed and corrected with current control (visual feedback).

What is interesting to note is that many of these early studies were concerned with improving human performance in the workplace or in connection with the work of the military. The next part of this section outlines some of the landmark studies from the early work to that conducted in the last few years.

Key players and motor control landmarks

The work of Woodworth (1899) has already been covered, but a number of other studies have been conducted that have formed the basic foundation of research in the area of motor control. A number of the key researchers and their studies are reviewed here in chronological order.

Charles Sherrington (1906)
Work on reflexes and sensory receptors by Sherrington formed the foundation of a number of major concepts in motor control. He was the first to conduct studies on sensory receptors and he was the first to use the term proprioception.

Edward Thorndike (1914)
Thorndike was interested in learning and he developed the term 'law of effect', which basically states that responses that are rewarded tend to be repeated. This theory still influences teaching and coaching today and is still studied by psychologists.

Franklin Henry (1939)
It is Franklin Henry who did the most to advance the study of motor behavior in relation to physical education and movement studies. He developed a whole new approach to research which combined psychological techniques with empirical laboratory-style experiments. Based in the Physical Education department of Berkeley, California, he is considered to be the founder of motor behavior research.

William Hick (1952)
Hick looked at the relationship between reaction time and the number of response alternatives. He found that choice reaction time is linearly related to the log of the number of stimulus–response alternatives. Therefore Hick's law states that reaction time (RT) increases by a linear amount (about 150 ms) each time the number of response alternatives doubles.

Paul Fitts (1954)
Fitts conducted an experiment in which subjects moved a stylus back and forth between two targets as quickly as possible. The distance between the targets and the width of the targets was varied. Fitts found that the relationship between movement time, distance between the targets, and target width was linear. This resulted in the development of Fitts' law, an equation linking the distance traveled, size of object and movement time. Fitts found that the relationship between the amplitude (A) of the movement, the target width (W), and the resulting movement time (MT) was given by the following equation:

$$\mathrm{MT} = a + b\,[\log_2 (2A/W)]$$

where MT is the average movement time, and a and b are constants.

Nicolai Bernstein (1967)
The writings of Bernstein have had a huge impact on the field of motor control, and his work received most attention some 20 years after it was written, once it was translated into English. Bernstein raised a number of questions about the control and coordination of movements. He tried to characterize the study of movements in terms of the problems of co-coordinating and controlling a complex system of biokinematic links. In particular, he worked on the issue of degrees of freedom.

Jack Adams (1971)
Jack Adams developed the 'closed loop theory' of control. This theory has been a great landmark in learning theory. Regardless of whether or not it is correct, it has inspired a great deal of research into motor learning.

Research Highlight: John Whiting (1929–2001)

The term 'keep your eye on the ball' is often used in sport and by the media. This phrase was first used by John Whiting who conducted a number of experiments on catching, with the emphasis on the importance of input variables on performance. Working in the Department of Physical Education at the University of Leeds he produced the seminal book *Acquiring Ball Skill*, in 1969 (Whiting, 1969). The studies conducted at Leeds on catching found the following:

- it is unnecessary to view the entire trajectory of a ball
- only 60 ms are available to close the fingers around the ball when it is moving at 10 m/s
- fingers started to close before the ball touched the hand.

These studies were ahead of their time and, interestingly, Whiting focused on the dynamic task of catching when most other researchers focused on discrete non-dynamic tasks. In 1975 he started the *Journal of Human Movement Studies*, further adding to the legacy and contribution to sports science.

Richard Schmidt (1975)

In 1975 Schmidt published a paper proposing a new theory of learning. Schema theory was a challenge to the work of Adams, and this new theory was a catalyst for a whole host of learning studies. An important part of Schmidt's work was the idea of a generalized motor program (GMP) which addresses the storage issue when learning new skills. Richard Schmidt was also the founder of the *Journal of Motor Behaviour* in 1969. This journal is, to this day, one of the leading journals in the field.

James Gibson (1979)

Gibson introduced the ecological approach to control and learning which emphasizes the interrelationship between movement and perception. He suggested that the environment has 'affordances' or 'invariant properties', which allow us to perceive directly the potential uses of objects. Gibson's work has resulted in a vast range of research that examines perceptual variables and the control of actions.

Why study motor control?

As motor control is concerned with the study of movement, it is vital that any individual who is concerned with coaching, or teaching sport or physical education, or involved in rehabilitation has a good grasp of all areas within the field of motor control. An understanding of the execution of all processes that lead to skilled human movement as well as factors leading to the breakdown of such skills is needed. When studying, we are asking questions about control and coordination that have implications for sport, activity and rehabilitation (Feldman and Meijer, 1999). Understanding the processes involved in performing and learning motor skills will therefore enable practitioners to:

- plan and deliver appropriate curriculum materials
- conduct more effective and targeted practice and rehabilitation experiences.

A practitioner is then better placed to plan or develop procedures to assist motor learning or motor development, or to assist the performer in enhancing motor control. All coaches, teachers and clinicians must ensure that any work they conduct is set with a framework of a sound theoretical starting point. An understanding of motor control is therefore vital.

Conclusion

This section provides the starting point to the rest of the book by defining key terms which will be used throughout the remaining sections. In addition some of the many landmark studies and key researchers have been introduced; much of their work will be considered in greater detail in other sections of the text.

Further reading

Elliott, D. E., Helsen, W. F. and Chua, R. (2001) A century later: Woodworth's (1899) two-component model of goal-directed aiming. *Psycholl Bull* **127**, 342–357.

Latash, M.L. (2001) *Classics in Movement Science.* Human Kinetics, Champaign, IL.

Savelsbergh, G. and Davids, K. (2002) 'Keeping the eye on the ball': the legacy of John Whiting (1929–2001) in sport science. *J Sports Sci* **20**, 79–82.

Swinnen, S. (1994) Motor control. *Encyclopedia of Human Behavior*. Academic Press, New York, vol 3, pp.229–243.

Thorndike, E. (1932) *The Fundamentals of Learning*. Teachers' College Press, New York.

Whiting, H.T.A. (1969) *Acquiring Ball Skill: a psychological interpretation.* Bell and Sons Ltd, York.

References

Adams, J.A. (1971) A closed loop theory of motor learning. *J Mot Behav* **3**, 111–149.

Bernstein, N. (1967) *The Co-ordination and Regulation of Movement.* Pergamon Press, London.

Bowditch, H.P. and Southard W.F. (1882) A comparison of sight and touch. *J Physiol* **3**, 232–254.

Bryan, W.L. and Harter, N. (1899) Studies on the telegraphic language: the acquisition of a hierarchy of habits. *Psychol Rev* **6**, 345–375.

Feldman, A.G. and Meijer O.G. (1999) Discovering the right questions in motor control: movements. *Motor Control* **3**, 105–135.

Fitts, P.M. (1954) The information capacity of the human motor system in controlling the amplitude of movement. *J Exp Psychol* **47**, 381–391.

Gibson, J.J. (1979) *The Ecological Approach to Visual Perception.* Houghton-Mifflin, Boston.

Hick, W.E. (1952) On the rate of gain of information. *Q J Exp Psychol* **4**, 11–26.

Hill, L.B., Rejall, A.E. and Thorndike, E.L. (1913) Practice in the case of typewriting. *Pedag Semin* **20**, 516–529.

Keogh, J.F. and Sugden, D.A. (1985) *Movement Skill Development*. Macmillan, New York.

Schmidt, R.A. (1975) A schema theory of discrete motor skill learning. *Psychol Rev* **82**, 225–260.

Sherrington, C.S. (1906) *Integrative Action of the Nervous System*. Scriber, New York.

Whiting, H.T.A (1969) *Acquiring Ball Skill: a psychological interpretation*. Bell and Sons Ltd, York.

Wolpert, D.M., Ghahramani, Z. and Flanagan, J.R. (2001) Perspectives and problems in motor learning. *Trends Cogn Sci* **11**, 487–494.

Woodworth, R.S. (1899). The accuracy of voluntary movements. *Psychol Rev Monograph Supplement* **3(3)**.

Section B

CLASSIFICATION OF SKILL

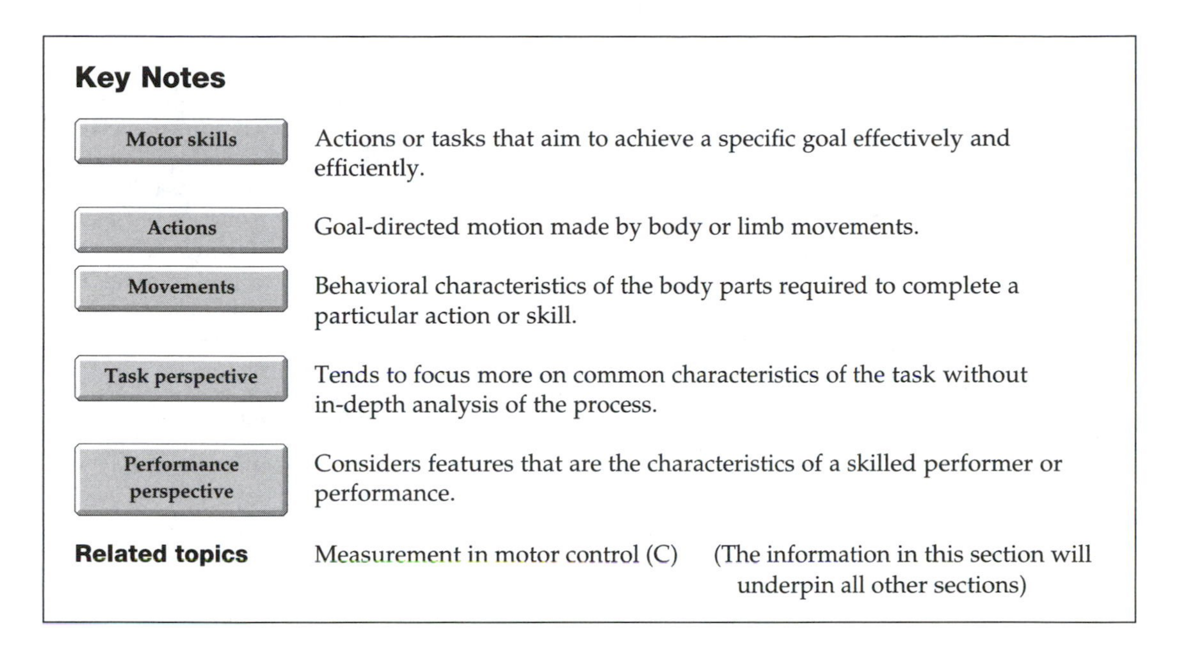

Key Notes

Motor skills	Actions or tasks that aim to achieve a specific goal effectively and efficiently.
Actions	Goal-directed motion made by body or limb movements.
Movements	Behavioral characteristics of the body parts required to complete a particular action or skill.
Task perspective	Tends to focus more on common characteristics of the task without in-depth analysis of the process.
Performance perspective	Considers features that are the characteristics of a skilled performer or performance.
Related topics	Measurement in motor control (C) (The information in this section will underpin all other sections)

A good starting point when considering control and coordination is to think about how to define movement and describe movements, and to reflect on how they are classified. Defining movement enables us to describe complex action with simple statements that are understood by coaches, teachers, clinicians, and performers. By using classifications we are better able to make generalizations and discuss principle from the same framework of understanding. Burton and Miller (1998, p.42) state that:

> Consistent definitions of key terms are a prerequisite for discussing and understanding movement.

It is therefore important to develop a framework of terms that we can use to describe and classify movement. Classification of movement patterns and skills provides further clues as to the nature of the anatomical, motor and mechanical requirements of a particular group of skills. This section aims to define terms that we use to describe movement, outline how we classify movement, and provide an overview of specialist terms that are used to summarize complex actions.

Definition of terms

When discussing or describing motor control there are many terms and classifications that are used in a specific way by researchers, coaches, teachers and therapists. We can all think of the terms that we generally use and that our teachers and coaches use, but do we have a common definition of them? It is important that we not only understand the terms used in motor control but also use them in the appropriate way. The first distinction that needs to be made is the difference between skills, actions, and movements. In motor control literature you will see the terms motor skills and movement skills used. These terms are often used to describe human performance but what do they actually mean? The list below defines skills, actions and movements.

- *Skills* (motor skills): actions or tasks that aim to achieve a specific goal efficiently.

The term skill is commonly used, and in our daily lives we perform a whole range of skills that vary in complexity. It must also be remembered that skills are learnt; how we learn and when we learn these skills will be discussed in Sections H and N. We must also remember that at times skills have to be relearnt due to injury or illness. In this text we are using the term motor skill to describe a task that has a specific goal, such as catching a ball. There are a number of characteristics that are common to motor skills. Firstly, motor skills have to be learnt and at times relearnt. Motor skills also have a purpose and are used to achieve or complete a task or goal. A motor skill also requires some voluntary body or limb movement. The term voluntary is important, as at times we produce reflex actions but these are very different from voluntary motor skills.

- *Actions*: goal-directed movements that consist of body/limb movements.

The term action is generally used for a particular motor skill. For example, you may say 'the action of bowling'. This may cause some confusion but we can assume that the term action and motor skill are often used in the same context. Keogh and Sugden (1985, p.7) state that action is 'concerned with the movements intent and function and its interplay with the environment'. Taking account of the interaction between the action to be performed and the movement context provides a richer overview of the nature of movement production. This will be explored in detail when we consider two-dimensional classifications of skill.

- *Movements*: the behavioral characteristics of the body parts required to complete a particular action or skill.

Movements are the component parts of skill and the term movement indicates the use of a specific body part or a combination of limbs. We link together a range of movements to complete a skill. Movements are therefore the behavior demonstrated when performing a particular skill in a particular context. However, the same skill can be completed or achieved via different combinations of movements. For example, we throw a ball differently on a windy day compared to a still day but the movement on both days is still considered throwing. This movement flexibility will be explored further when we look at theories of control. Movement is often described in functional terms, such as throwing or catching. This is very different from describing the body movements that are produced to achieve a functional outcome. Movements are also adjusted to the movement context; in *Fig. 1* the kayaker makes adjustments in line with changes in environmental conditions. As we consider the classification of movement we will begin to see the importance of looking in-depth at the nature of movement and movement production.

The above terms are used to describe activities in contexts where humans are moving. They are not intended to describe the quality of the skill, action or movement produced. Across the literature and depending upon the discipline, you will find differences in how the above terms are applied. Burton and Miller (1998) comment on the importance of being consistent in the terms that we use to describe movements but also state that terms such as movement skill and motor skill are often used synonymously and that this can cause confusion. As it

Fig. 1. When paddling a kayak the performer has to make many adjustments due to environmental demands.

is likely that differences in the terminology used by different disciplines will continue to vary, they suggest that the way forward is to agree on consistent definitions for key terms. In addition, they recommend that a taxonomy of movement skills should be developed that takes into account the needs of different disciplines.

Classification systems have been developed to further help describe and evaluate skills, actions and movements. These tend to begin with the term skill to make the first distinction, and use definitions of the actions and/or movements to add further detail to the classifications. Classification systems that have been developed tend to approach motor skill from two perspectives: a task perspective or a performance proficiency perspective. Fleshman *et al.* (1984) state that classifications are the arrangement of objects into groups based on their properties. The remainder of the section will outline the most popular methods of classifying motor skill.

Task perspective skill classifications

Task perspective skill classifications tend to focus more on what is actually happening without in-depth analysis of the process. Four distinctions have been used to classify tasks from a task perspective:

1. the way the task is organized
2. the importance of the motor and cognitive elements
3. the level of muscular involvement
4. the level of environmental predictability.

Some of these classifications are one-dimensional and just take into account one of the above on a linear continuum, others are two-dimensional (Gentile, 1987) and take into account more than one of the above and consider how they interact.

Skills classified by task organization

Skills are often classified by task organization with discrete skill and continuous skills at opposite ends of the continuum. *Table 1* outlines such a classification giving examples from sport and physical activity.

Table 1. Discrete, serial and continuous skills

Discrete	Serial	Continuous
Kicking	Pole vaulting	Running
Throwing	Long jump	Swimming

Discrete skills are usually brief, with a well-defined beginning and end. Discrete skills are often used in motor control experiments because the experimenter can easily identify the start and the finish of the movement. In reaching to grasp an object, for example, the movement starts when the hand begins to move and ends when the object is grasped. Other examples of discrete skills include hitting, kicking, throwing, and catching.

Serial skills involve several discrete actions with the order of the action being crucial to success. For example, in long jumping there is the run-up, the take-off, the flight and the landing. These sub-elements have to be performed in a specific order for the jump to be successful. Serial skills can be considered as a number of discrete skills linked together. Serial skills are sometimes learnt most effectively by dividing them into their component parts, such as when learning a gymnastic routine. By their very nature serial skills usually have a longer movement time than discrete skills.

At the other end of the continuum are *continuous skills* that are often repetitive with no recognizable beginning or end. Examples of these are running, cycling and swimming, which of course do end at some point but during the event the performer is continually moving. Continuous skills are often repetitive, rhythmic and involve a long time span. The duration of a continuous skill is usually determined by a barrier or finishing line.

Skill classified by the importance of the motor and cognitive elements

Skills are also classified on a continuum with the main determinant of success being purely the movement at one end (motor skill) and quality of the decision or the cognitive component at the other end (cognitive skill). As can be seen in *Table 2*, some skills demand a combination of both. In fact, it could be argued that some level of cognitive involvement is required to perform most skills.

Table 2. The continuum from motor skill to cognitive skill

Motor skill		Cognitive skill
Motor skill maximum	Both motor and cognitive skill	Cognitive skill maximum
Weight lifting	Rock climbing	Sports coaching
Shot putt	Playing hockey	Playing bridge

This is very much a continuum and the focus towards the more motor or the more cognitive element of a task or skill may well depend upon the level of experience or expertise of the performer. For example, when you are first learning to bowl in cricket you will focus on the motor elements of the task. As you become more skilled you will be able to focus more on tactics. It is also interesting to note that when skills have a high cognitive component, such as driving a car, we are, after extensive practice, eventually able to reduce this task to mainly a motor activity. We can then introduce other cognitive tasks which we are able to perform at the same time without having to pay much attention to the motor component of the task.

The level of muscular involvement

Skills are also classified by level of muscular involvement. Again, this is a one-dimensional classification with gross motor skills at one end and fine motor skills at the other (*Fig. 2*).

Gross ———————————————— Fine

Fig. 2. The continuum of motor skills from gross to fine.

Gross motor skills refer to movements involving large muscle groups, such as weight lifting. Fine motor skills refer to more precise skills involving smaller muscle groups, such as manipulation tasks. As with other one-dimensional classifications we have to remember that movement is complex and such scales are often too simplistic to fully explain movement. Gross and fine skills are often performed at the same time, for example if we are sitting typing we need manual dexterity (fine motor skills) to work the keys but gross motor skills to maintain posture to enable us to sit at the desk.

Skill classified by the level of environmental predictability

Another way to classify motor skill is to consider the extent to which the environment is stable and predictable. Again this is a way of classifying skill based on a one-dimensional continuum. At one end are open skills where the environment is variable and unpredictable such as in most team games. At the opposite end of this are closed skills where the environment is stable and predictable as in performing a front one-and-one-half somersault in springboard. When swimming or running 400 m the distance remains the same and the environment is relatively stable and predictable. However, because wind conditions may vary and the performance of others is somewhat unpredictable, these events may still be thought of as situated between closed and open.

Two-dimensional classifications and taxonomies

Over the past 50 years many movement taxonomies have been developed. Taxonomies, as we shall see later, have had an impact on assessment – especially those that have taken into account the structure of motor abilities and the classification of motor skills. Here we will focus on Gentile's taxonomy, which has been influential both in classifying movement and in developing assessment tools. So far the classification systems covered have taken a one-dimensional approach to defining skill. Such classification systems do not fully account for the complex nature of movement or provide robust enough frameworks from which to describe or assess movement.

Gentile (1987) developed a two-dimensional classification system or taxonomy that enables a more precise description of movement. This classification system combines the requirements of the action and the demands of the environment. Gentile's classification is referred to as a taxonomy of movement skills that subdivides movement into 16 categories based on four task factors with two levels in each (*Table 3*).

- *Taxonomy*: a classification system that is organized according to the relationships among the component characteristics to be classified.

Classifying skill in this way provides a more realistic view of movement in the context in which it is performed. The development of this taxonomy was designed for use by physiotherapists as it enables comprehensive assessment of the patient's movement characteristics, and this provides the basis for selecting appropriate activities for rehabilitation. However, this taxonomy is useful for any individual who is involved in assessing movement and can be used for teaching and coaching as well as rehabilitation.

Gentile has refined and developed her classification system since 1975, with further elaborations in 1987 and 2000. As can be seen in *Table 3*, the taxonomy has two major movement descriptors that can be subdivided into 16 movement categories. The complexity of the movement increases as you move across and down the table. Action function forms one of the dimensions that Gentile's taxonomy is based upon. Gentile takes the function of an action and begins to classify movement by deciding if the body (person) needs to be stationary or moving to perform the task, and whether performing the task requires an object to be manipulated or not. Many tasks that we perform do not involve body

Table 3. Gentile's classification of skill showing the components of the 16 categories

			ACTION FUNCTION			
			Body orientation stable		**Body orientation moving**	
			No object manipulation	*Object manipulation*	*No object manipulation*	*Object manipulation*
ENVIRONMENTAL CONTEXT	**Regulatory condition: stationary**	*No inter-trial variability*	1 Watching a football game	2 Writing at a desk	3 Walking or running	4 Ten-pin bowling
		Inter-trial variability	5 Using sign language	6 Shooting in archery	7 Performing a step-up in the gym	8 Running over an obstacle course carrying a ball
	Regulatory condition: moving	*No inter-trial variabilty*	9 Standing in a lift or on an escalator	10 Standing and bouncing a ball	11 Walking on a moving walkway at the airport	12 Dribbling a football around a set of cones
		Inter-trial variabilty	13 Standing on one leg when wearing roller skates	14 Catching balls thrown at different speeds	15 Crossing the road	16 Playing rugby, hockey, cricket etc

movement; in fact they require the body to be stable in order to perform the task. Examples of this are sitting, standing and watching sporting events. Thus, the first part of the classification establishes if the task requires the person to be moving or stationary and the next element involves considering if the task involves any manipulation of objects. Gentile subdivides each element of body orientation into no object manipulation or object manipulation. If the person's body is stable when performing a task do they have to manipulate any objects? If you were playing chess your body would be stable but you would have to manipulate the pieces. Such an activity would be placed in box 2 in Gentile's taxonomy. Ten-pin bowling involves moving the body and manipulating the ball. Therefore the body would be moving and there would be object manipulation (box 4). We perform many daily tasks that involve us moving and manipulating objects.

It must be remembered that a task becomes more complex when we are moving and manipulating an object. Not only do we have to control our arm to position our hand or hands correctly (this will alter drastically depending upon the need for power or the need for dexterity), but we also have to maintain appropriate postural control.

While discussing the above we have not taken into account the context of the environment. The second aspect of Gentile's taxonomy does take account of the context of the environment in which the movement is taking place. Gentile refers to this as environmental context and this is first subdivided into regulatory conditions. The regulatory conditions are characteristics of the environment that regulate movement. Regulatory conditions can be either stationary (stable) or moving. When we are walking or running or hitting a golf ball off the tee, the regulatory conditions are stationary or we could say the environment is stable. However, in cricket when we hit or field the ball the environment is moving and the body may be stable or moving. Movements where both the body and the environment are moving can be very complex and are therefore more difficult to perform. Gentile's taxonomy further subdivides regulatory conditions into inter-trial variability and no inter-trial variability. When there is no inter-trial variability the conditions influencing the task would be the same each time the person repeated the task. For example, practicing shooting in basketball or netball would be the same each time the person performed the task. However, placing a defender on the court would provide inter-trial variability as they would probably attempt different tactics each time to stop you from scoring. When playing games such as hockey or football the regulatory conditions would therefore be moving and there would be inter-trial variability making the environmental context complex. This would also involve a complex action function with the body moving and object manipulation taking place. Box 16 on Gentile's taxonomy contains tasks that are the most complex; examples of these are often found in the sporting domain. We must also remember that tasks that we take for granted can be more complex than they first appear; walking is one example. Patients with spinal cord injury have to relearn how to walk. The complexity of this task and the role of the environment can be seen, as once they have learnt to walk in the clinic they also have to learn to walk in the outside world where the regulatory conditions are not as stable. The 16 boxes of Gentile's taxonomy represent a gradual increase in task complexity from the completely closed skills in box 1 where the environment and body are stable and stationary with no inter-trial variability, to the open skills of box 16 where the environment and the body are moving with inter-trial variability.

Activity: Classification of skills

At the times listed in the table below make a note of what you are doing (for example at 10.00 am you might be walking to a lecture). Then classify the skill from a task organization perspective and by the level of environmental predictability.

Task	Skills classified by task organization	Skills classified by the level of environmental predictability
e.g. 10.00 am walking	Continuous	Unpredicatable
10.00 am		
2.00 pm		
5.00 pm		
8.00 pm		

Now consider how the tasks you were performing during the day could be classified using Gentile's (1987) taxonomy of motor skills.

There are many advantages in using Gentile's taxonomy as a foundation for assessing movement skill. It enables a full evaluation of a person's movement ability in a variety of movement contexts. This is important for both rehabilitation and teaching and learning. More importantly, the taxonomy assists in the planning of functionally appropriate activities that increase in difficulty as learning occurs or the person is rehabilitated. Such is the usefulness of Gentile's taxonomy that other researchers have used her framework as a starting point for assessing and classifying movement.

Research Highlight: Henderson and Sugden (1992)

The Movement Assessement Battery for Children (Movement ABC) developed by Henderson and Sugden (1992) also incorporates a framework or task classification element that takes into account the individual and the environment. As in Gentile's taxonomy consideration is given to the state of the individual and environment. Henderson and Sugden's task classification takes into account whether the child is stationary or moving and if the environment is stable or changing. This forms the basis for a number of tasks that are performed within the Movement ABC that vary in difficulty in order to identify children with developmental coordination disorder. This demonstrates how flexible and appropriate it is to use a two-dimensional classification system in order to assess and describe movement in a variety of settings.

The Movement ABC gives a performance proficiency score on a range of performance tasks. Children are tested on manual dexterity, static/dynamic balance, and ball skills that vary in difficulty depending upon whether the child/environment is stable or moving. Performances are given a score between 0 and 5 on each item so the total score can vary between 0 and 40. The raw scores are recorded on the score sheet and are matched with the corresponding scaled score using a six-point scoring system. Following completion of the test, the item scores are totalled and expressed as a percentile rank. A higher score on the tests indicates a higher level of impairment. The scores are then compared to age norms and the child can then be placed on a continuum of percentiles, below the 5th percentile indicating the child needs immediate remediative action and below the 15th percentile indicating the child to be at risk. A comment is also made by the tester on any other physical or behavioral influences on performance. Other performance proficiency assessments are also available and these will be discussed later as, unlike the Movement ABC, they are not based on classification of skills but are purely tools for assessing particular tasks such as walking or catching.

The next section looks at classifications of movement from a performance proficiency perspective.

Classification from a performance-proficiency perspective

A starting point for considering performance in terms of proficiency is to look at features that are the characteristics of a skilled performer or performance. Schmidt and Wrisberg (2000) view proficiency in terms of: maximum certainty of goal achievement, minimum energy expenditure, and minimum movement time. This encompasses the definition of skill proposed by Guthrie (1952). However, in the section we are going to consider broader classifications or taxonomies that have been developed to describe performance proficiency (*Fig. 3*).

- When performing a skill we can examine both the process and the outcome.

Performance proficiency is a general term for achieving or completing the task as quickly as possible with as little effort as possible. Our ability to do this can be measured by a number of outcome measures that will be discussed in the next section. By concentrating on the above we are still getting a limited view of the performance and the processes involved. We know that movement is complex and diverse and that the classification of skill is an important starting point for movement assessment. Burton and Miller (1998) stress the importance of movement assessment to a range of professionals including sports scientists.

Fig. 3. When performing a skill we can examine both the process and the outcome.

They state that there is a need for 'consistent definition of key terms' and the development of more multidisciplinary-based taxonomies of movement skill. A number of movement skill assessment tools have been developed recently that have a more functional definition of movement skill. From a classification starting point, that takes into account the task, the environment and the individual, and therefore enables a more accurate assessment of performance (for a full review see Burton and Miller 1998). Many of these classifications or taxonomies take into account both outcome and process measures (see Research Highlight, p.16).

Other classifications have taken a more specific focus to movement. A classification of manipulative hand movements has been developed by Elliott and Connolly (1984). This classification deals with intrinsic movements which involve the manipulation of an object within and by the hand, not movements which involve the transportation of the hand by the arm. Proficiency can also be measured by making an assessment of how an individual completes a given task.

Elliott and Connolly (1984) have categorized these movements into three classes: simple synergies, reciprocal synergies and sequential patterns, each of which has a range of movement patterns within it (*Fig. 4*). The categories of movements used are functionally based and include most manipulative movements. The first distinction made between movements is whether they are simultaneous or sequential. In simultaneous movements the digits are used together in a single movement pattern, in sequential movements there is an

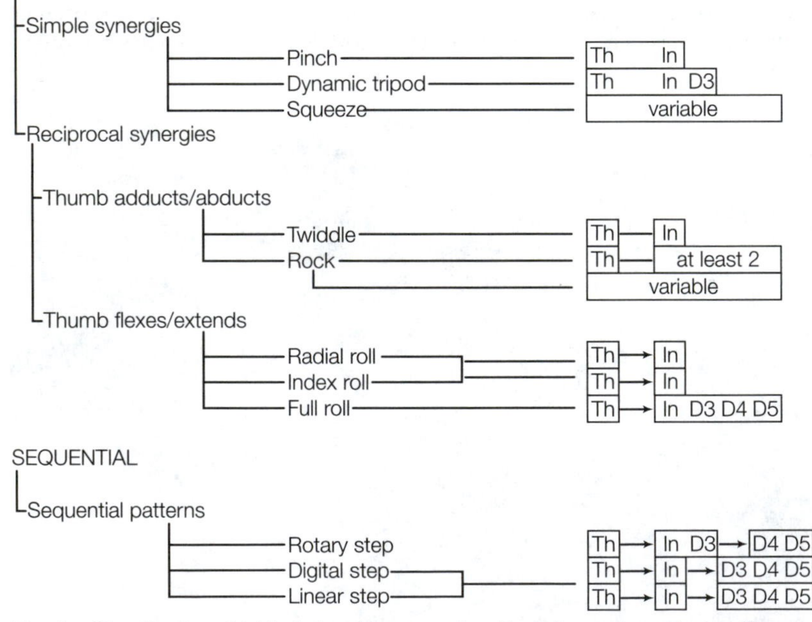

Fig. 4. Classification of intrinsic hand movements with digits employed in specific movement patterns. Each box represents one or more digits; if more than one, they move together in the same way. Linking boxes by simple lines indicates reciprocal action between groups defined by boxes; linking by arrows represents sequential actions. Redrawn with permission from Elliott and Connolly 1984.

involvement of the digits in an independent manner. The first two categories of simple and reciprocal synergies both involve simultaneous movement patterns. Simple synergies (pinch, dynamic tripod and squeeze) are those that involve all the participating digits, including the thumb, in convergent flexor/extensor synergies, such as in squeezing a rubber ball.

Reciprocal synergies (twiddle and rock), in comparison, involve movements in which the thumb and participating digits perform dissimilar movements, for example the flexion of the fingers and extension of the thumb that occurs in the fine manipulation of objects, such as rolling a piece of plasticine between the thumb and first two fingers. The third category, sequential patterns (rotary step, digital step, and linear step) involves a discontinuous movement of the object with some digits supporting the object while others reposition it, which in turn become the supporters while the former are involved in repositioning the object. Turning a pencil around in the fingers is an example.

When considering reaching and grasping, the above classification demonstrates the variety of movements and grasps that the hand can perform. It also reminds us that how the arm moves and which grasp is employed is governed by the size, shape, and position of the object. This provides a comprehensive classification of manual skills enabling detailed descriptions of the nature of the performance of manual skills made by an individual.

As already stated, these classifications help us describe movement from a set framework; however we also know movement is complex, and more diverse classifications are probably needed to fully account for the nature of the movement required. For example, it is difficult to perform a fine skill such as drawing a thin line on paper without postural control, which is provided by large muscle groups and could be considered a more gross skill. Two-dimensional classifications are better able to account for the complexity of movement as they consider the environment and the function of the action.

Movement terminology

This section finishes with some definitions of a range of specialist terms that are used in motor control and development to describe movement. These terms will be used throughout this text and are listed here as a reference. The first part defines terms that define types of movement, the second outlines key anatomical or movement terms that are used when describing movement; this includes definitions of terms that describe the nature or the number of limbs or body parts involved in movement production.

Types of movement

- *Abduction*: the movement of the limb away from the midline of the body.
- *Adduction*: the movement of the limb towards the midline of the body.
- *Associated movements*: movements made in addition to the movement being observed. For example, when running a person may make excessive head movements.
- *Circumduction*: the movement of the limb in a circular motion at the distal end.
- *Dorsiflexion*: the movement of the foot towards the leg.
- *Eversion*: the sole of the foot faces laterally such that the big toe is lower than the little toe.
- *Inversion*: the sole of the foot faces medially such that the little toe is lower than the big toe.
- *Extension*: occurs when a limb or body part is moved so that the angle

between the two limbs or body parts is increased. An example of extension would be moving the arm backwards from the shoulder. Extension is the opposite of flexion.

- *Flexion*: occurs when a limb or body part is bent so the two limbs or body parts come closer together. Such movements usually occur in the sagittal plane, for example bending forward at the waist.
- *Plantar flexion*: movement of the foot away from the leg.
- *Rotation*: the turning movement of a limb around its long axis.
- *Pronation*: movement of the forearm where the palm is turned outwards and the radius and ulna are parallel.
- *Supination*: movement of the forearm where the palm is turned inwards and the radius rotates over the ulna. (Some clinicians will use the term 'pronation and supination' to represent movement about the foot during stance.)

Key anatomical or movement terms

- *Anterior vs. posterior*: front vs. back
- *Cranial vs. caudal*: head vs. tail
- *Dorsal vs. ventral*: back vs. front
- *Extrinsic*: an external constraint such as task demand or contextual influences
- *Intrinsic*: an internal constraint such as limb length or limb dominance
- *Ipsilateral vs. contralateral*: same side vs. opposite side
- *Medial vs. lateral*: toward the midline vs. away from the midline
- *Proximal vs. distal*: close to the torso vs. further from the torso
- *Superior vs. inferior*: above vs. below (*Fig. 5*)
- *Unimanual*: involving one arm (*unilateral*: involving one side)
- *Bimanual*: involving both arms (*bilateral*: involving both sides)
- *Tetra*: involving three limbs
- *Quad*: involving four limbs
- *Ventral*: toward the front
- *Dorsal*: toward the back
- *Rostral*: toward the nose
- *Caudal*: toward the tail
- *Superior*: toward the top of the head/body
- *Proximal*: closer to the origin of a structure
- *Distal*: further away from the origin of a structure.

It is important that these terms are used when we describe movements as they are universally understood and enable complex actions to be explained by a few simple phrases or words.

Frame of reference

Movement of the body takes places around planes and axes, and using these terms assists our understanding of the nature of the movement produced.

Understanding planes will facilitate learning and the use of terms related to position of structures relative to each other and movement of various parts of the body. *Figure 6* shows the three planes of the body about which movement takes place. The frontal or coronal plane separates the body into anterior (front) and posterior (back). The sagittal or median plane separates the body into right and left parts, and the horizontal or transverse plane separates the body into superior (top) and inferior (bottom).

Anatomical axes are like rods or lines that pass through the body. They are

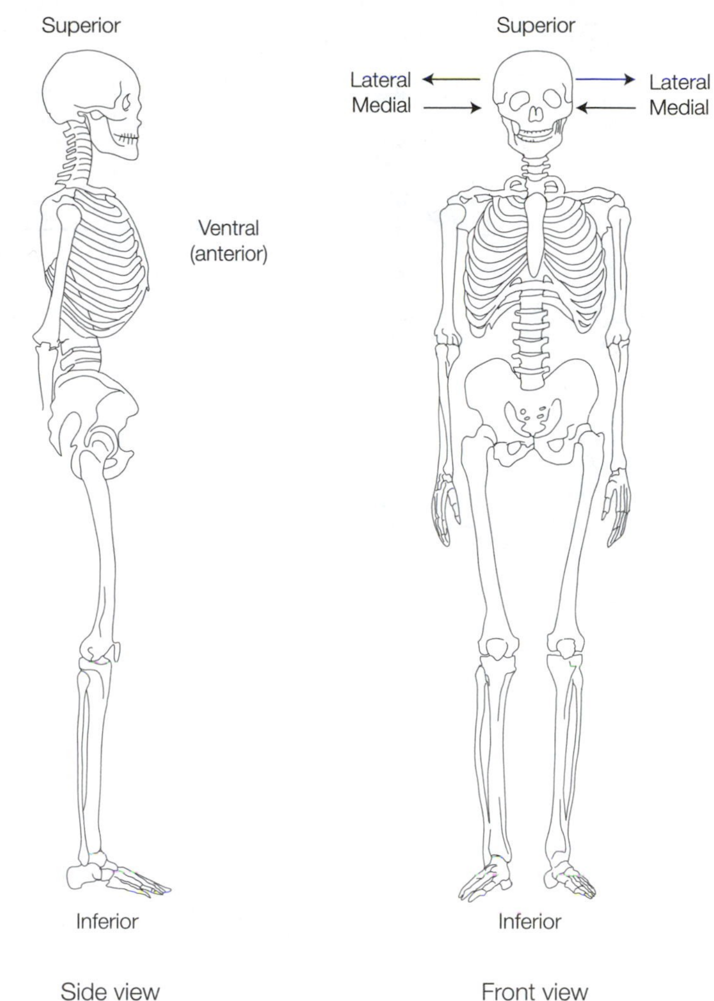

Fig. 5. Anatomical terms.

used to describe axes about which rotational movements take place. The longitudinal axis passes from the top to the bottom as if a rod had been placed in the head. The horizontal axis passes from left to right and the anterior–posterior axis from front to back.

Conclusion

The classifications of movement and the terms used in this section will be used throughout this book. You will also find that an understanding of these terms better enables you to describe movement, plan for intervention (teaching and coaching) and engage with other texts and articles that are concerned with movement, control, learning and development.

Further reading

Palisano, R.J., Cameron, D., Rosenbaum, P.L., Walter, S.D. and Russell, D. (2006) The Manual Ability Classification System (MACS) for children with cerebral palsy: scale development and evidence of validity and reliability. *Dev Med Child Neurol* **48**, 549–554.

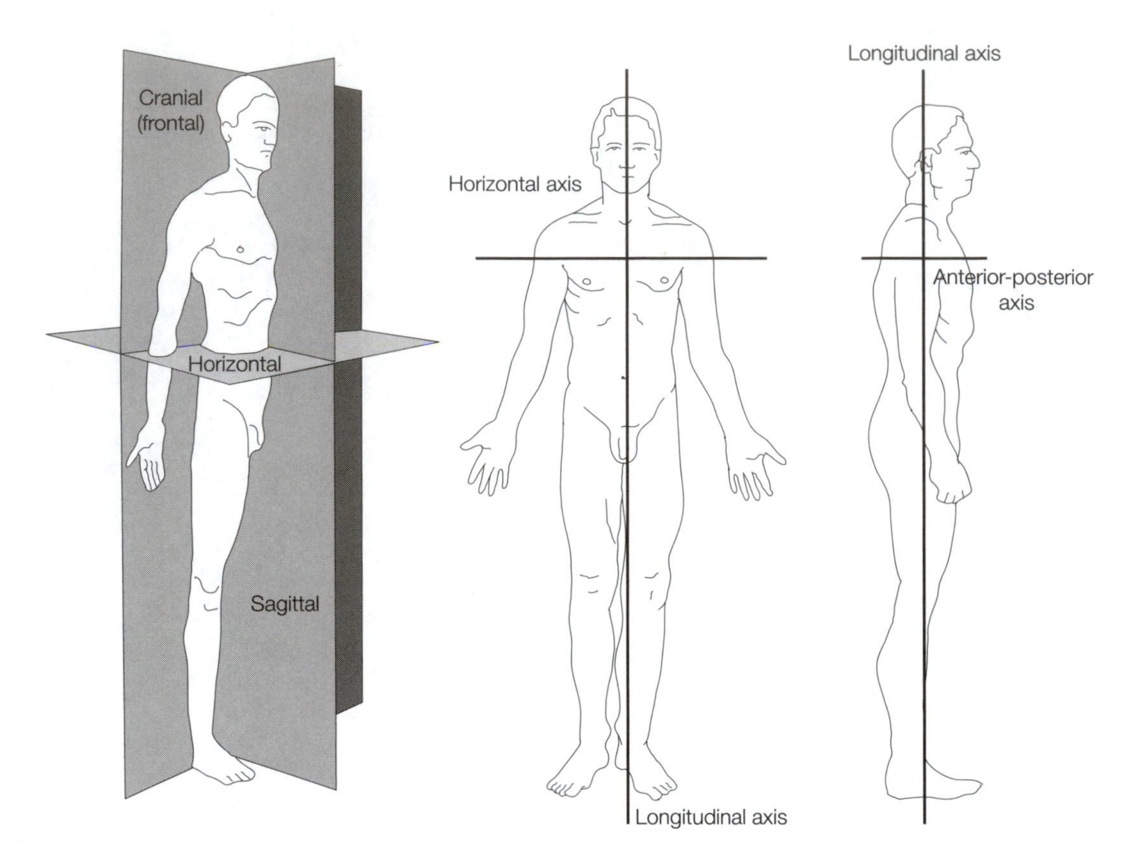

Fig. 6. Planes (left) and axes (right) of the body.

Singer, R.N. and Chen, D. (1994) A classification scheme for cognitive strategies: implications for learning and teaching psychomotor skills. *Res Q Exerc Sport* **65**, 143–151.

References

Burton, A.W. and Miller, D.E. (1998) *Movement Skill Assessment*. Human Kinetics, Champaign, IL.

Elliott, J.M. and Connolly, K.J. (1984) A classification of manipulative hand movements. *Dev Med Child Neurol* **26**, 283–296.

Fleshman, E.A., Quaintance, M.K. and Broedling, L.A. (1984) *Taxonomies of Human Performance*. Academic Press, Orlando, FL.

Gentile, A.M. (1987) Skill acquisition: action, movement, and neuromotor processes. In: Carr, J.H. and Shepherd, R.B. (eds) *Movement Science: foundations for physical therapy in rehabilitation.* Aspen Publishers, Rockville, MD, pp. 93–154.

Guthrie, E.R. (1952) *The Psychology of Learning: revised edition*. Harper Bros, Massachusetts.

Henderson, S.E. and Sugden, D.A. (1992) *Movement Assessment Battery for Children.* The Psychological Corporation, Sidcup.

Keogh, J.F. and Sugden, D.A. (1985) *Movement Skill Development*. Macmillan, New York.

Schmidt, R.A. and Wrisberg, C.A. (2000) *Motor Learning and Performance*. Human Kinetics, Champaign, IL.

MEASUREMENT IN MOTOR CONTROL

Key Notes

Outcome measures	Measures that indicate the outcome of performing a particular skill. Usually measured in time, distance, height, or some form of numerical measurement.
Performance measures	Measures that indicate the details or specific characteristics of how a skill or action is produced.
Learning curves	Graphical representations of improvements in performance.
Validity	The extent to which a test or measure measurers what it is designed to measure.
Reliability	The consistency or repeatability of test scores or data collected.

Related topics

Theories of motor learning (H)
Implications for practice (K)
Motor development (M)

Whatever the context, measurement forms an important part of motor control. Measures may provide an indication of control, a way of looking at learning, or an assessment of development or recovery. Measurement is evidence and it enables us to compare, contrast, assess, and re-evaluate all aspects of movement. The coach, teacher or therapist will take measurements that inform their practice and provide a variety of information. There are many ways in which we can measure performance or movement. Some measures require very little in terms of equipment, others are very sophisticated and require a high level of technology. Throughout this book we refer to measures and consider research that uses a variety of measures. It is vital therefore that we have an understanding of the nature of measures taken in motor control and understand issues of validity and reliability. Taking measurements in motor control serves a number of purposes. As already stated, measurement is very much concerned with assessment, identification, and the evaluation of movement. The following list provides some examples of the role of measurement.

1. *Planning*: measurement enables the coach/teacher/therapist to find a starting point (baseline) and to plan future sessions.
2. *Feedback*: measurement provides feedback both for the coach/teacher/therapist and also the performer or patient.
3. *Evaluation or diagnosis*: a performer's level of expertise or level/extent of injury can be assessed by measurement. This enables the action/movement to be described in detail.

4. *Prediction*: measurement can be used to predict or to assess models of control, learning and development.
5. *Comparisons*: measurement enables the movements of others to be compared or the movement of a particular performer to be compared before and after some intervention or in different contexts.
6. *Monitor*: by taking measurements we can assess progress and development.

Before taking measurements we have to consider the classification of the skill, the main purpose of the skill and the component parts of the movement to be performed. (Section B provides an overview of skill classification.) The primary purpose of the skill, action or movement to be performed may well be one of the following:

- *balance*:
  - to gain stability or postural control
  - to regain stability
  - to maintain stability
- *locomotion*:
  - move from point to point
  - move a prescribed distance
  - move using a prescribed pattern
- *projection (of self or an object)*:
  - for maximum height
  - for maximum accuracy
- *manipulation*:
  - of objects
  - of another person
- *self-help*:
  - dressing
  - feeding.

Once this is established we have to discover or calculate what are the component parts of the task? It may be that the following components have to be taken into account:

- joints and muscles involved
- coordination or timing issues
- neuromuscular considerations
- objects to be used
- environmental context
- related inputs.

The measurement of all of the above can take place at many levels. Generally, measurement can be divided into two areas; firstly, outcome measures and secondly, performance measures. Both can be a rich source of information but at differing levels. These measurements include:

- basic time taken or score achieved
- non-invasive psychological measures
- experimental procedures
- experimental procedures that monitor or alter CNS function
- neurological measures.

The remainder of this section will outline outcome and performance measures used in motor control before considering issues of validity and reliability.

Outcome measures

Included in this section are measures that indicate the outcome or result of performing a motor skill. The most common that is used is time. The time taken to complete a task or the length of time the performer is able to perform a task for is often measured. As can be seen in *Table 1*, this could be time taken to run a mile, time in balance, or time in contact. For the therapist it may be time taken for a patient to walk a set distance after injury. In a study examining mode of therapy for patients with spinal cord injury (Sitthikongsak, 2005), the participants were asked to walk as far as they could and the distance covered was measured pre-intervention, post-intervention and at weeks 2 and 4 during rehabilitation (*Fig. 1*). Here the outcome measure of distance covered was used to assess patients and to compare two methods of rehabilitation.

Time to respond to a stimulus is also a common motor control measurement. Many researchers have been interested in reaction time and choice reaction time. This issue will be dealt with in depth in Section E.

Another common outcome measure when looking at performance is magnitude. Magnitude is often used in the sporting context as it considers distance,

Table 1. Common outcome measures

Outcome measure	Example
Time	
Time to complete a task	Amount of time to run a mile
Time in balance	Time a performer can stay on a balance board or snowboard
Reaction time	Time between starter's gun (stimulus occurring) and movement response
Magnitude	
Distance	Long jump
Height	High jump
Angle	Degree of knee flexion
	Pronation/supination
Accuracy	
Amount of error	Number of centimeters away from target
Number of percentage errors	Number of free throws missed
Number of successful attempts	Number of times the bean bag hit the target
	Number of catches

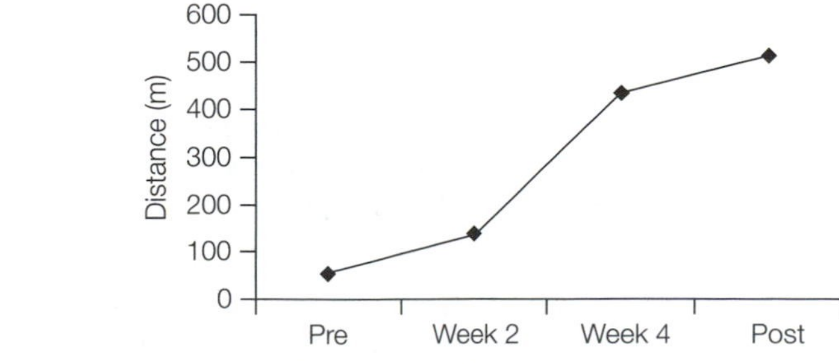

Fig. 1. Mean distance walked by participants with spinal cord injury.

height, and angle. Not only does this enable comparisons to be made between athletes for example, but it is an excellent method of comparing a patient's functional recovery during rehabilitation. A therapist may measure the degree of flexion/extension as an indicator of recovery after a shoulder injury. Another outcome measure that is often used is accuracy. The measurement of accuracy can be applied at a basic level, such as whether the ball was caught or dropped, or it can be dealt with at a complex level. Especially in the experimental setting accuracy could refer to maintenance of speed or force at a designated level. It could also be how accurate a performer is in archery or curling; here the degree or level of accuracy could be measured. When long jumping, take-off could be measured in terms of accuracy. On one level it could be if the performer hit the board, or at another level it could be how many millimeters away from the line the athlete was.

Error measures

When measuring sport, activity and performance we are often interested in the amount of error that has occurred. Measures of error provide information about accuracy (by how much did the penalty taker miss?) and they also provide information about performance. Three main calculations made are absolute error, constant error, and variable error. Absolute error is the amount or magnitude of error made without reference to the direction that the performer has deviated from the target. Absolute error (AE) therefore provides information about the performer on a trial or a set of trials. Constant error (CE) is a measure of the amount and direction of error and therefore provides more information about the nature of the performance. For example, does the footballer tend to kick the ball too high when taking a penalty and by how much? Another method of assessing performance is to consider consistency. This is done by calculating variable error (VE), which is determined by calculating the standard deviation of a performer's constant error score. These error measurements can be very useful when looking at accuracy and consistency which is an important aspect of sporting performance.

Remember that performance outcome measures do not tell the performer anything about the behavior of the limbs that led to the outcome. Outcome measures therefore tell us what has happened but they do not tell us why. As can be seen in the Research Highlight in this section (p.34), more in-depth information is needed to fully understand the performance characteristics.

Performance measures

By examining the performance components we are looking in detail at the mode by which the outcomes are produced. A whole host of performance measures are available that vary in terms of the depth and complexity of the information that they provide. In the next part of this section we provide an overview of the performance measures most commonly used. These measures can tell us about:

- control and coordination
- the functioning of the nervous/sensory system
- the functioning of the muscular system
- the actions of the limbs or joints
- the interaction between the above systems.

As we will see later when we look at coaching and practice considerations, information about performance can be provided before, during and after a person has performed the skill. As already stated there are a range of methods that can be employed that enable us to gain performance measures. Some of these methods require relatively simple and easily available equipment. Others require advanced and expensive technology.

Video analysis

Video analysis is a readily available tool for the researcher, coach or clinician. By using video analysis we can gain a mass of information about movement. This can range from the analysis of games such as football where we might be interested in the number of passes made, the amount of time spent active, or the number of passes completed (Dawson *et al.*, 2004). In contrast we may use frame-by-frame analysis to examine gait in patients recovering from injury, to examine human performance or to examine components of control (Bradshaw and Aisbett, 2006; Glaister *et al.*, 2007). Often researchers use video analysis to assess movement in relation to a scale or assessment battery. Sugden and Utley (1995) used video analysis to examine reaching and grasping unimanually and bimanually in children with cerebral palsy. They examined the movement frame by frame and made observations on changes in trajectory, changes in posture, and associated movements. Catching has also been examined using video analysis with researchers interested in a range of constraints in both one-handed and two-handed catching (Bennett *et al.*, 2002; Astill and Utley, 2006). Video analysis is therefore an excellent tool for the examination of performance and it is able to produce data that are very revealing on all aspects of control and coordination. Previous work examining catching behavior in children highlighted the need for a sensitive catching scale for children, as often a 'clean' catch is not made (Bennett *et al.*, 1999). Therefore it was decided to score each experimental trial on a 0–5 point scale developed especially for use with young children and adapted for assessing a two-handed catch (see Wickstrom, 1983, p,163; *Table* 2). The misses were classified as those trials in which the child's hand completely missed the ball. On the 0–5 point scoring scale, a complete miss would be awarded a 2, 1 or 0. Such scales can be used in conjunction with video as filming the movement and then applying an assessment criterion enables more reliable measurement to be made.

The use of video analysis should not be underestimated as it is a tool that is more likely to be available and can provide both outcome and performance measures.

Table 2. Description of catching behavior in relation to points scoring scale of Wickstrom (1983, p.163)

Points scored	Description of catching behavior
5	Clean catch: ball is contacted by the catching hands only and retained
4	Assisted catch: ball is juggled initially by the catching hands and then retained
3	Hand contact: initial contact with the hand but the ball is subsequently dropped
2	Upper body contact only and the ball is dropped
1	Lower body contact only and the ball is dropped
0	No ball contact with a part of the body

Kinematic methods

Kinematic measures are often used to examine human performance. The use of the term kinematic indicates the description of movement without reference to force or mass. Kinematic measures are taken by marking the body segments of interest with markers which can vary in size and shape, and then recording the movement of the markers in one, two, or three dimensions. Once filmed the path of these markers is then calculated either by manually entering them into a computer by a variety of software packages or by automatic detection and digitization. *Figure 2* and *3* show examples of cameras and markers often used in kinematic analysis.

Much of the literature in this area is concerned with the latest developments in software and hardware and also on the merits of various camera positions, film speeds, method of marking subjects, and type of reference scales used. It must be remembered that accuracy is an important aspect of data collection and the researcher has to ensure that the data collected are as valid and reliable as possible (Borghese *et al.*, 1997; Challis *et al.*, 1997).

> . . . skin markers will reflect the movements of the underlying bone comparatively well. On the other hand, where the soft tissues are thick (such as at the hip) or mobile in relation to the bone (such as over the scapula) skin markers will tend to only reflect skin movement. (Lungberg, 1996, p.412)

Fig. 2. Participant climbing with 3D skin markers.

Fig. 3. 3D camera system.

Many researchers interested in human movement use kinematic measures, and the Further reading section (pp.38–39) suggests a number of studies that will be of interest to those involved in sport, activity and rehabilitation. When undertaking kinematic analysis, data that are commonly collected are velocity, displacement and acceleration measurements. Velocity indicates the rate of change of an object or person in relation to time, displacement indicates the limb or joint changes in position, and acceleration indicates the change in velocity over time. *Figure 4* provides an example of this, taken from displacement data from a reach-and-grasp action. Other examples can be seen in the Research Highlight in this section (p.34). A range of equipment is available for kinematic analysis including systems by Pro reflex, Vicon, Qualisys, and MIE Leeds.

Kinetic methods

Kinetic measures take into account force and mass. This includes both internal and external forces. Various types of equipment are available to take such measurements, including force plates and force transducers. Researchers are usually interested in ground reaction forces (*Fig. 5* and *6*) and joint torque, which is concerned with the forces of rotation at the joint. These types of measures are especially useful to the clinician and those involved in gait research (Tenore *et al.*, 2006), but many studies involve investigation into force in the sporting context (Hill, 2002; Manoogian *et al.*, 2006; Stiles and Dixon, 2006).

As can be seen in *Fig. 5* and *6*, the ground reaction forces in barefoot sprinting are much higher than those in barefoot running and contact is longer in running

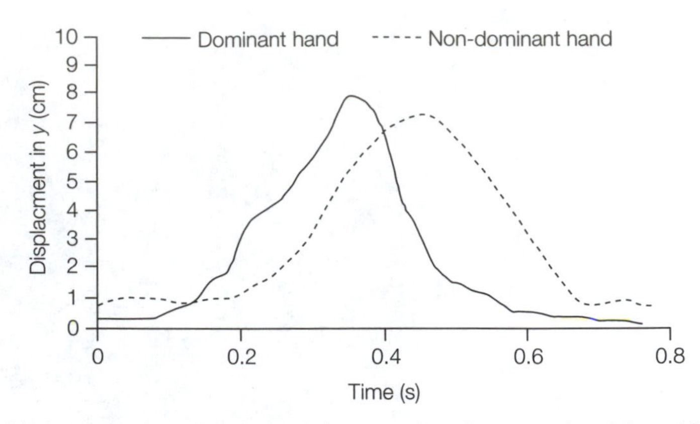

Fig. 4. Displacement data of the wrist during a reach-and-grasp action. Adapted from Utley et al. *(2007).*

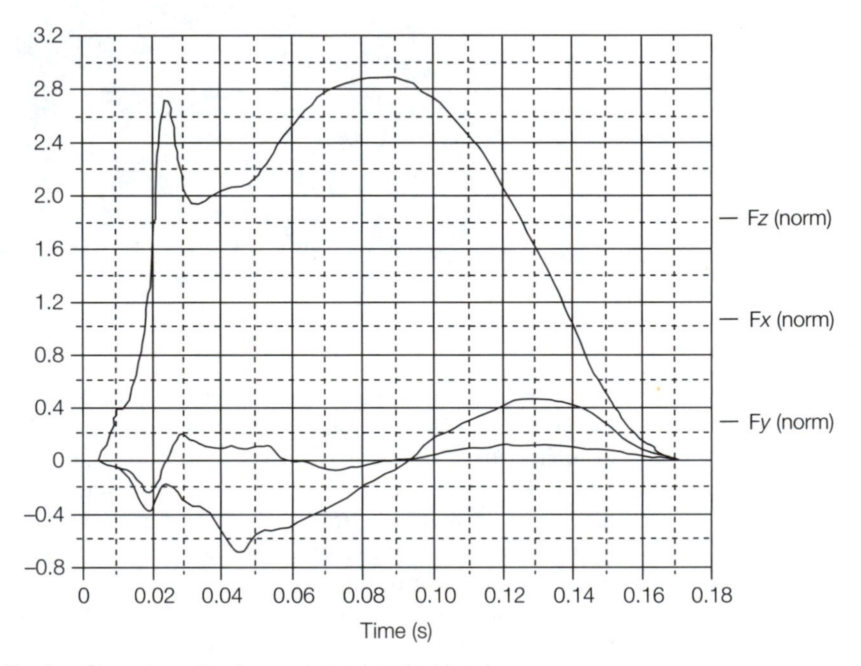

Fig. 5. Ground reaction forces during barefoot jogging.

than in sprinting. Such data enables a range of comparisons both within and across participants to be made. In addition different activities can be compared and the clinical applications are numerous.

Electromyography (EMG)

When moving, electrical activity takes place in the muscles which can be measured using electromyography (EMG). EMG provides data that are especially useful to those studying motor control. This includes information about muscle activation and also, by examining groups of muscles, provides information about coordination. At the end of the eighteenth century, Galvani

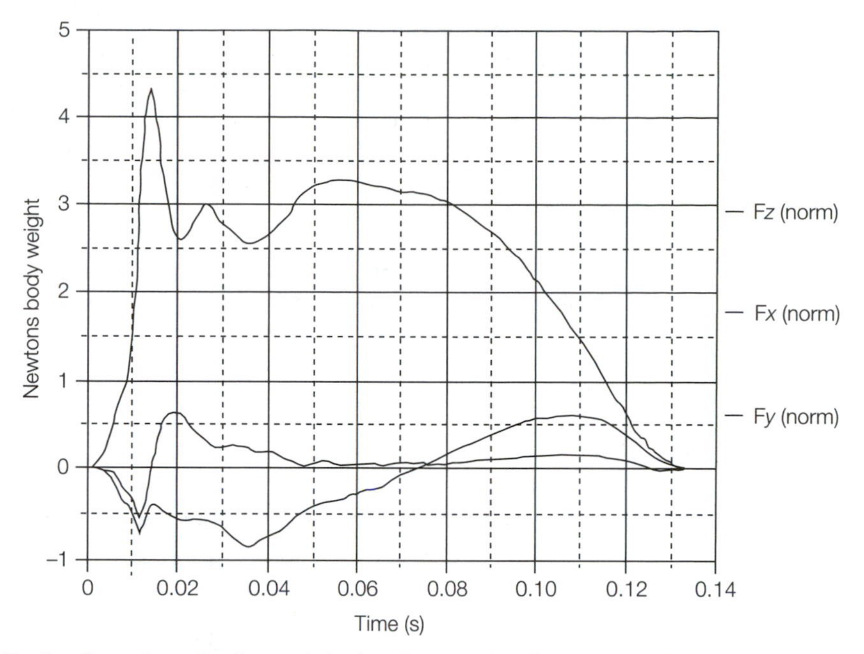

Fig. 6. Ground reaction forces during barefoot sprinting. Fx, Fy and Fz, force in the x, y and z axes.

discovered electromyography (EMG) in experiments where he depolarized the muscles of frogs' legs by touching them with metal rods. However, it was not until the 1960s that EMG measurement was used to study human movement. More recently, advances in technology and a greater understanding of the electrical properties of the body have allowed more reliable methods of measuring EMG to be developed. Measuring EMG activity involves the use of electrodes; there are three main types of electrodes, surface electrodes, needle electrodes and wire electrodes. When using electrodes the right electrode for a specific function should be selected (see below), and it is vital to identify and place the electrode on the correct muscle.

Surface electrodes are used for:

- time–force relationship of EMG signal (quantified EMG after Spaepen)
- kinesiological studies of surface muscles and large muscle groups
- neurophysiological studies of surface muscles
- psycho-physiological studies
- interfacing muscle with an external device such as electro-myo artificial limbs.

Needle electrodes are used for:

- study of control properties of motor units (firing rates, recruitment etc)
- exploratory clinical electromyography (nerve conduction velocity studies).

Wire electrodes are used for:

- kinesiological studies of deep muscles
- neurophysiological studies of deep muscles
- kinesiological studies of small or adjacent muscles where cross-talk may occur.

Neurological measures

In addition to EMG, neurological measures such as electroencephalography (EEG) and functional magnetic resonance imaging (fMRI) are now used in motor control research. These methods enable us to understand the function of the brain and the central nervous system (CNS) to a depth that previously could only be inferred from behavior. EEG and fMRI are used in motor control in order to understand brain organization during voluntary movement. Functional magnetic resonance imaging is a technique for establishing which parts of the brain are activated during activity or movement. This is achieved using an MRI scanner so that the increased blood flow to the activated areas of the brain shows up on fMRI scans. In EEG, electrodes are placed on the scalp over multiple areas of the brain in order to detect and record patterns of electrical activity and check for abnormalities. EEG enables us to investigate in detail the temporal characteristics of brain activity during isometric contractions or movements, allowing us to examine changes in temporal processing of a movement. EEG's advantage over fMRI is that it does not constrain the type of activity or level of exertions that are difficult if not impossible to perform in the enclosed environment of the fMRI scanner.

Eye tracking

As researchers have become interested in all aspects of control and coordination, new technology has enabled equipment such as eye trackers to be developed. The most widely used eye trackers are video based and they enable the movement of one or both eyes to be monitored. Most modern eye trackers use contrast to locate the center of the pupil and use infrared and near-infrared non-colluminated light to create a corneal reflection. A camera focuses on one or both eyes and records the eye movement as the performer attends to a stimulus. The camera records very rapid eye movements and eye movement associated with scanning the environment. There are various eye-tracking systems, some are head-mounted, some require the head to be stable, and some function remotely and automatically track the head during motion.

The participant in *Fig. 7* is using an ASL Model H6 eye-tracking system which is lightweight with head-mounted optics that allow unrestricted freedom of movement. This system is portable and therefore allows the participant to move around and interact with the environment as they normally would. The optics are mounted on an adjustable headband and the scene being viewed by the participant is recorded with a color camera. The images from the eye can then be collected and calculations made as to where the participant was looking during the recording. This has great implications for both the sporting and clinical setting. Experiments have examined eye tracking in a range of activities from visual search strategies employed by gymnastic coaches with different levels of expertise (Moreno *et al.*, 2002), to studies looking at gaze of low- and higher-handicap golfers (Vickers, 1992). In both these studies, expert participants showed longer and fewer visual fixations than the novice group. Other studies have examined how sports performers use visual information to anticipate the direction of a moving object. Moreno *et al.* (2005) looked at 12 participants with different competitive experience in sports when performing an interception task. A tennis ball was fired at participants from a ball machine; the spot where the balls bounced was analyzed by a digital camera and visual behavior was examined by an eye-tracking system. Analysis showed that the non-experienced participants made significantly more errors and were more variable in the

Fig. 7. Participant using a drive simulator where eye movements are recorded.

visual-occlusion conditions. As this technology improves we will gain a better understanding of the role of vision in controlling movement.

Measuring learning

A range of methods can be employed to measure or record performance, and from these it can be inferred that learning has taken place. Methods that can be used include:

- performance curves
- retention tests (assessed before and after training)
- transfer tests (look at how skill transfers to a new situation/skill).

When considering skill development/learning, there are four performance characteristics that indicate that improvement or learning is taking place. These are improvement, consistency, persistence and adaptability. Over time, a performer should show improvement and be able to perform at a higher level. Therefore the performer should become more stable or consistent in performing the skill. Finally the true measure of learning taking place is that the skill learnt is adaptable and can be demonstrated in a range of contexts. In Section K we will consider how practice can influence each of these.

Performance curves

Improvements in performance can be observed from performance curves, with some examples shown in *Fig. 8* and *9*. Typical features of performance curves are:

- rapid initial improvement and more gradual improvement later
- one or several plateaus in performance where no gains are made.

However, performance curves tend to follow four general trends which can be seen in *Fig. 8*. A linear curve indicates that performance improved at a constant

Research Highlight: Astill and Utley (2006)

A study on catching by Astill and Utley looked at both outcome and performance measures in children with and without developmental coordination disorder (DCD). The outcome measure (number of successful catches) indicated that on average the DCD group caught statistically significantly fewer balls than the age-matched controls (AMC) group (DCD = 12 ± 3; AMC = 23 ± 1). In order to discover why, performance measures have to be used. In the figure below, the velocity of a typical catch from each group reveals performances differences. The outcome is therefore very different for each group.

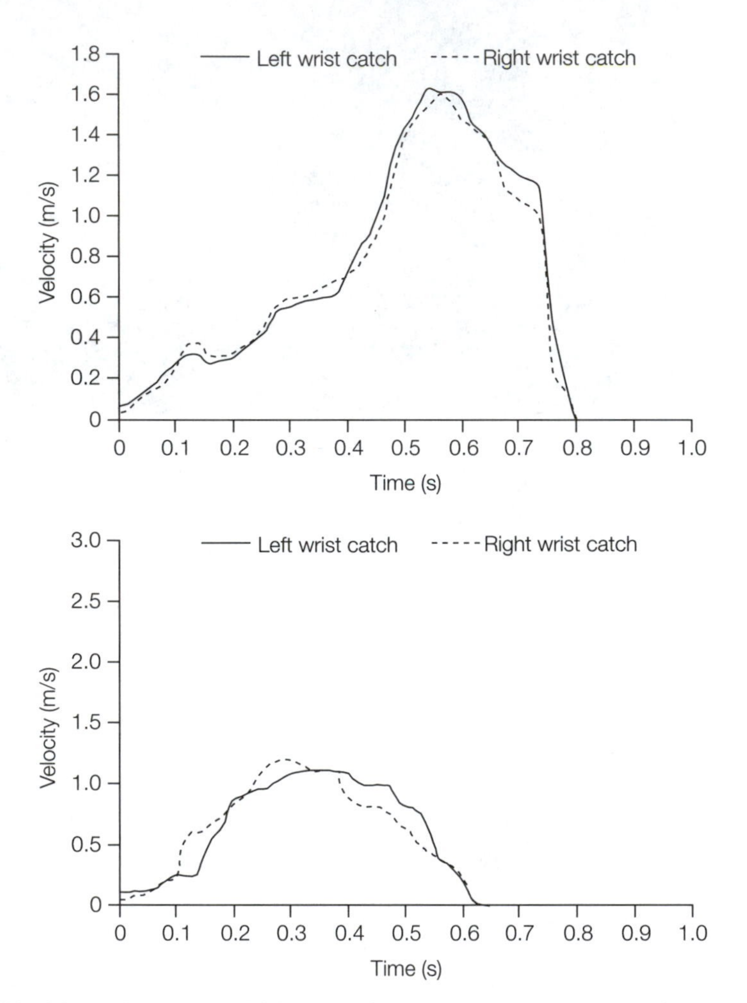

Velocity profiles of the left and right wrists in DCD children (top) and age-matched controls (bottom). Reproduced with permission from Astill, S. and Utley, A. (2006). Two-handed catching in children with developmental coordination disorder. Motor Control **10**, *520–532.*

Velocity profiles provide information on the performance, which explains the difference in outcome. The bell-shaped profile of the AMC children is shorter in duration and has a lower peak velocity and a shorter acceleration phase when compared to that of the children with DCD. In addition, the limbs do not seem to be as strongly linked, indicating a more advanced catching action. This work demonstrates the need to use both outcome and performance measures.

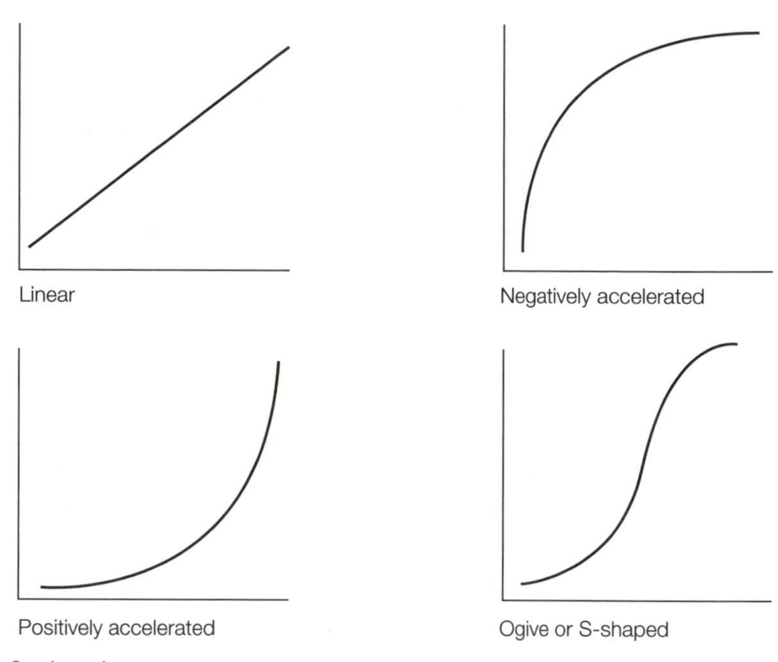

Fig. 8. Learning curves.

rate over the trials and that between each trial the same amount of improvement occurred. If a negatively accelerated curve is produced, this indicates that performance improvements were rapid in the early trials and then slowed down as the trial went on. In a positively accelerated curve there is little improvement initially, but performance improvements are much greater in later trials. Finally, in an S-shaped curve performance improved slowly at first, then very quickly and then slowly again.

Curves will vary depending on the nature of the measures taken. They may show a gradual decrease if they display error rates or reaction times. There can also be problems with showing group results because the differences between individuals are masked; also they smooth the look of the curve across individuals, whereas the progress of each individual may be in jumps. *Figure 9* shows the mean results from 20 participants who took part in a balance experiment. Participants had to balance on a stability board for 30 s, and the time they spent out of balance was recorded. As can be seen, learning took place as participants improved from over 20 s out of balance to less than 10 s out of balance over 25 trials.

Retention tests

Retention tests measure how much you know or have retained from studying or practicing. The normal method of using these is to assess performance with a criterion task pre-intervention or practice and then post-intervention or practice. This will give an indication of performance increase or improvement. However, if you are interested in learning you should administer the test again, after a period where no practice has taken place. If the score is higher than the baseline (or lower depending upon the task), then you can assume that some learning has

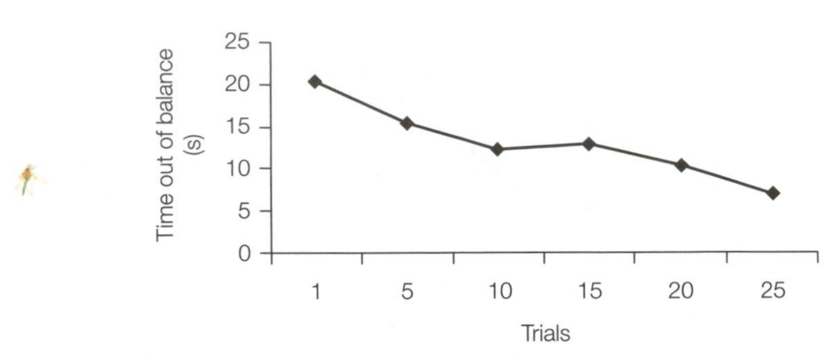

Fig. 9. Time out of balance on a dynamic balance task.

taken place. Exactly how much is more difficult to determine, but employing this method does at least give an objective basis for assessing that learning has or has not occurred. This type of test is commonly used in motor skills research. A number of variations could be used but the general format would be:

Pre-test → intervention → post-test retention tests.

In *Fig. 10*, improvement in catching was assessed by evaluating the flexibility or adaptive nature of the coordination system when catching a ball one-handed. The results indicated that participants post-practice were able to move their limbs at a faster rate, and that this continued after practice had ceased indicating that a higher skill level has been attained. The participants could move their limbs faster and with a greater degree of flexibility in line with Bernstein's stages of learning (see Section H).

Transfer tests

Transfer tests are tests that assess how the learner is able to apply the skill that has been learnt in a different context. Transfer tests are especially important when learning open skills where the real-world situation is likely to be slightly different each time (e.g. tennis, football). The skill of dribbling a football in practice has at some point to be transferred to the game situation. Transfer tests are also referred to as novel tasks where the individual attempts the skill learnt but some component of the task is modified. A person may learn to juggle with bean bags but then

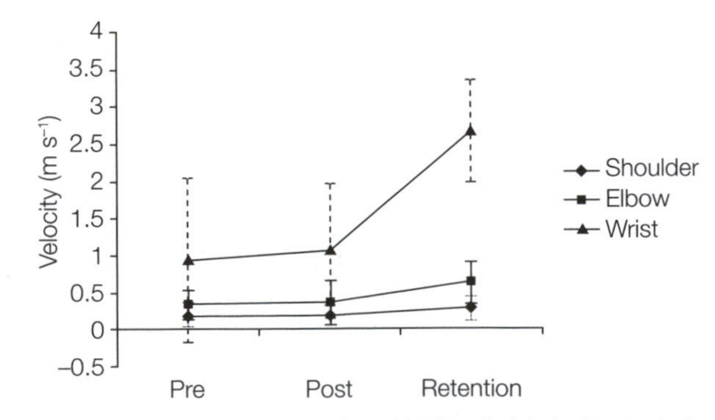

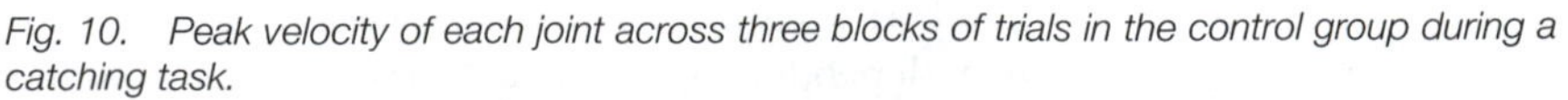
Fig. 10. Peak velocity of each joint across three blocks of trials in the control group during a catching task.

attempts to juggle with tennis balls. The context, skill or both can be manipulated to make the task novel. Transfer tests are important in terms of assessing learning and performance, and any experiment examining learning should consider using transfer tests.

Performance plateaus

Performance plateaus occur when performance which had been improving suddenly reaches a steady state with little or no improvement, after which improvements start again. These performance plateaus seem to be experienced in real life, but researchers have had difficulty in demonstrating them in experiments. One question researchers have asked is whether or not plateaus are common in learning, or whether they are a phenomenon of practice. Plateaus are more often identified in the results of individuals, not groups. In addition, performance curves tend to mask plateaus because mean scores smooth the individual effect. In *Fig. 9* there appears to be a plateau between trials 10 and 15. A number of explanations for plateaus have been proposed. It may be that plateaus represent a period of transition, a period of low motivation or fatigue, or they could be due to the measure employed causing a ceiling effect.

Validity and reliability

When taking measures or developing measures in motor control it is vital that we ensure that the data collected are valid and reliable. Here we are concerned with issues of accuracy and consistency.

Validity

Validity refers to the accuracy of a measure. A measurement is valid when it measures what it is supposed to measure and performs the functions that it purports to perform. When taking measurements in motor control we have to ask whether an indicator accurately measures the variable that it is intended to measure. There is a range of approaches to validity that are used to establish or demonstrate the accuracy of a measure or test. Some involve the use of statistics, some make a comparison to a standard, and others are qualitative.

Content validity refers to the degree or extent to which the questions or task measures the task or desired information. The content of the measure or instrument used to gather the information is established by expert judgments on the appropriateness of the content of a test. This then becomes the 'gold standard' that is used. This type of validity is not considered to be strong.

Face validity is determined by judgments made by the researcher and is based on surface appearance. For example, a 20 m sprint may be taken as a test of speed but we do not know how accurate this is. Is this really a measure of speed? Face validity is therefore the weakest type of validity.

In concurrent validity, the test or measure would be validated by comparing scores taken within a close period of time. The test therefore needs to correlate with an already-validated measure. The accuracy of the test would be determined by the degree of statistical relationship. If I score a player's tennis ability by having them complete a battery of tests on Monday, I would repeat this on Tuesday to ensure that the results were a true reflection of their ability.

A construct stands for a collection of related behaviors that are associated in a meaningful way. Construct validity is used when the variable to be measured has no definitive criterion and is therefore difficult to measure. If we want to devise a measure that identifies potential problems with eating disorders we would develop an instrument and then give this to distinctly different groups of people, some of whom may have had eating disorders in the past. We would test if the instrument could identify any difference between the groups. If it did we could assume that the measure had construct validity.

Reliability

Reliability refers to how consistently and dependably measures are taken or data collected. It must be remembered here we are only concerned with scores or data not the instrument used. This is critical in research and there are a number of measures of reliability.

Inter-rater or inter-observer reliability is used to assess the degree to which different researchers give consistent estimates of the measure taken or movement observed. For example, if two researchers watch a video of a four-year-old child catching and assess the catch using the Wickstrom (1983) catching behavior scale would they get the same score? If they did, the score of catching for that child could be considered reliable.

Test–retest reliability is used to assess the consistency of a measure/score taken from one time to another. This is often used in motor control and learning. Here a measure would be taken with a group of participants, at one time, and later the same measure applied to the same participants again. You would then have two sets of scores and a correlation of the scores should find a high degree of association. Care must be taken not to administer the tests too close or too far apart otherwise factors such as fatigue or learning could influence the results.

Parallel-forms reliability is used to assess the consistency of the results of two tests constructed in the same way aiming to test the same variable. This is a form of equivalence reliability and a correlation of the scores should find a high degree of association. One difficulty of this method is developing two good tests that measure the same thing.

Internal consistency reliability is used to assess the consistency of results across items within a test. This is used to show how consistent the scores of the test are and is often used in performance tests that incorporate many test trials. A study on throwing might involve six tests that are administered in pairs. Reliability would be determined by comparing the results and correlating the paired tests.

Conclusion

Measurement is a crucial component of motor control, learning and development. This section has provided an overview of the types of measures taken in this area while taking into account issues of validity and reliability. In the remaining sections we will review a range of topics and research in which an understanding of measurement provides a key foundation.

Further reading

Andersen, T.E., Floerenes, T.W., Arnason, A. and Bahr, R. (2004) Video analysis of the mechanisms for ankle injuries in football. *Am J Sports Med* **32**, 69S–79S.

Ashford, D., Bennett, S.J. and Davids, K. (2006) Observational modeling effects for movement dynamics and movement outcome measures across differing task constraints: a meta-analysis. *J Mot Behav* **38**, 185–205.

Berg, K.E. and Latin, R.W. (1994) *Essentials of Research Methods in Health, Physical Education, Exercise Science, and Recreation*. Lippincott Williams and Wilkins, Philadelphia.

Davids, K., Lees, A. and Burwitz, L. (2000) Understanding and measuring coordination and control in kicking skills in soccer: implications for talent identification and skill acquisition. *J Sports Sci* **18**, 703–714.

Fiolkowski, P., Horodyski, M., Bishop, M., Williams, M. and Stylianou, L. (2006) Changes in gait kinematics and posture with the use of a front pack. *Ergonomics* **15**, 885–894.

References

Astill, S. and Utley, A. (2006) Two-handed catching in children with developmental coordination disorder. *Motor Control* **10**, 520–532.

Bennett, S., Button, C., Kingsbury, D. and Davids, K. (1999) Manipulating visual informational constraints during practice enhances the acquisition of catching skill in children. *Res Q Exerc Sport* **70**, 220–232.

Bennett, S.J., van der Kamp, J., Savelsbergh, G.J. and Davids, K. (2002) Discriminating the role of binocular information in the timing of a one-handed catch. The effects of telestereoscopic viewing and ball size. *Exp Brain Res* **135**, 341–347.

Borghese, N.A., Cerveri, P. and Ferrigno, G.C. (1997) Statistical comparison of DLT versus ILSSC in the calibration of a photogrammetric sterio-system. *J Biomech* **30**, 409–413.

Bradshaw, E.J. and Aisbett, B. (2006) Visual guidance during competition performance and run-through training in long jumping. *Sports Biomech* **5**, 1–14

Challis, J., Bartlett, R. and Yeadon, M. (1997) Image based motion analysis. In: Bartlett, R.M. (ed) *Biomechanical Analysis of Movement in Sport and Exercise.* BASES, Leeds, pp.7–30.

Dawson, B., Hopkinson, R., Appleby, B., Stewart, G. and Roberts, C. (2004) Player movement patterns and game activities in the Australian Football League. *J Sci Med Sport*, **7**, 278–291.

Glaister, B.C., Bernatz, G.C., Klute, G. and Orendurff, M.S. (2007) Video task analysis of turning during activities of daily living. *Gait Posture* **25**, 289–294.

Hill, H. (2002) Dynamics of coordination within elite rowing crews: evidence from force pattern analysis. *J Sports Sci* **20**, 101–117.

Lungberg, A. (1996) On the use of bone and skin markers in kinematics research. *Hum Mov Sci* **15**, 411–422.

Manoogian, S., McNeely, D., Duma, S., Brolinson, G. and Greenwald, R. (2006) Head acceleration is less than 10 percent of helmet acceleration in football impacts. *Biomed Sci Instrum* **42**, 383–388.

Moreno, F.J., Luis, V., Salgado, F., Garcia, J.A. and Reina, R. (2005). Visual behavior and perception of trajectories of moving objects with visual occlusion. *Percept Mot Skills* **101**, 13–20.

Moreno, F.J., Reina, R., Luis, V. and Sabido, R. (2002) Visual search strategies in experienced and inexperienced gymnastic coaches. *Percept Mot Skills* **95**, 901–902.

Sitthikongsak, S. (2005) *Effects of Treadmill and Conventional Walking Training in Patients with Incomplete Spinal Cord Injury (iSCI).* PhD thesis, University of Leeds.

Stiles, V.H. and Dixon, S.J. (2006) The influence of different playing surfaces on the biomechanics of a tennis running forehand foot plant. *J Appl Biomech* **22**, 14–24.

Sugden, D.A. and Utley, A. (1995) Vocabulary of grips in children with hemiplegic cerebral palsy. *Physiother Theory Pract* **11**, 67–79.

Tenore, N., Fortugno, F., Viola, F., Galli, M. and Giaquinto, S. (2006) Gait analysis as a reliable tool for rehabilitation of chronic hemiplegic patients. *Clin Exp Hypertens* **28**, 349–355.

Utley, A., Sugden, D.A., Lawrence, G. and Astill, S (2007). The influence of perturbing the working surface during reaching and grasping in children with hemiplegic cerebral palsy. *Disabil Rehabil* **29**, 79–89.
Vickers, J.N. (1992) Gaze control in putting. *Perception* **21**, 117–132
Wickstrom, R.L. (1983) *Fundamental Motor Patterns*, 3rd Edn. Lea and Febiger, Philadelphia.

Section D

THEORIES OF CONTROL

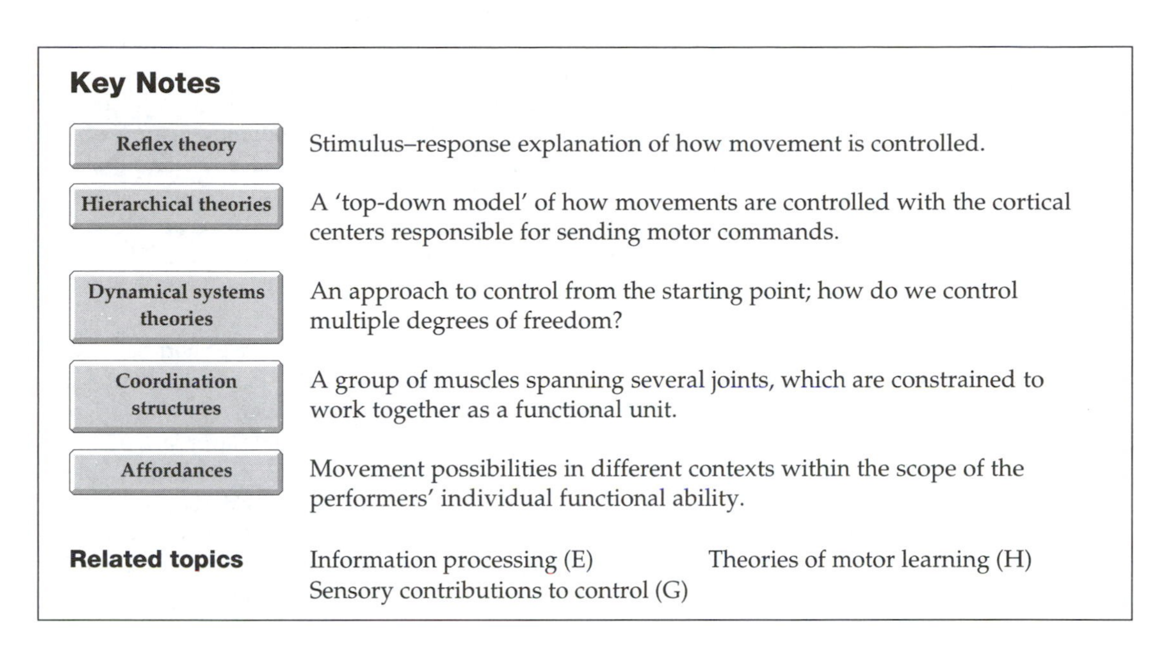

Key Notes

Reflex theory	Stimulus–response explanation of how movement is controlled.
Hierarchical theories	A 'top-down model' of how movements are controlled with the cortical centers responsible for sending motor commands.
Dynamical systems theories	An approach to control from the starting point; how do we control multiple degrees of freedom?
Coordination structures	A group of muscles spanning several joints, which are constrained to work together as a functional unit.
Affordances	Movement possibilities in different contexts within the scope of the performers' individual functional ability.
Related topics	Information processing (E) Sensory contributions to control (G) Theories of motor learning (H)

To successfully perform a wide variety of skilled actions the performer has to coordinate various muscles and joints to work both effectively and efficiently (*Fig. 1*) Each movement a performer makes has to be planned, constructed and executed. Various theories have evolved to explain the nature and control of movements.

Fig. 1. We can change our posture and muscular involvement to functionally achieve the same end result in different contexts.

Sheridan (1984) outlined four aspects of motor control that have to be considered:

> First, that the motor system has the ability to achieve functionally the same end result via different movements, involving different muscles and joints (motor equivalence). Second, that movements are never exactly repeated (uniqueness of action) . . . Third, that despite the apparent uniqueness of each movement the most obvious feature of skilled performance is the consistency and stability of both temporal and spatial structure (stability and consistency of action). Fourth, that not only is skilled action consistent and stable, but it is also capable of amendment as a consequence of changes in information available to the performer (modifiability of action). (Sheridan, 1984, p.60)

The variety of movements available to execute a task is massive. Sheridan's statement indicates the range of motor control problems that the performer has to cope with. Over the years there has been a change in emphasis from reflex theory to an information-processing approach and to a dynamical systems approach of movement control. It must be remembered that any theory of control must be able to account for motor equivalence, uniqueness of action, stability and consistency of action, and modifiability of action. This section will provide an overview of these theories and consider how we control complex movements (it is recommended that this section is read in conjunction with Section H). Theories of control will be considered under the following headings:

1. reflex theories
2. hierarchical theories
3. dynamical systems theories
4. ecological theories.

Reflex theories

Over the years there has been a change in emphasis from an information-processing approach to a dynamical systems approach to movement control. Prior to this, Sherrington (1906) proposed the reflex theory of control. After a series of excellent experiments, Sherrington concluded that with the whole nervous system, the reactions of various parts of that system, and the simple reflexes, are combined into greater actions that constitute the behavior of the individual as a whole (reflex chaining). Put in simple terms a stimulus creates a movement or reflexive response, see *Fig. 2*. There are a number of limitations to the reflex theory of motor control (Rosenbaum, 1991). It does not adequately explain and predict movement that occurs in the absence of a sensory stimulus. Furthermore, it cannot explain the production of fast movements or the ability to produce novel movements, or how a single stimulus can result in varying responses.

Hierarchical theories

As the term suggests, hierarchical theories of control work from a foundation that all aspects of movement planning and execution are the responsibility of higher orders of the central nervous system (CNS). In this top-down model, higher cortical centers send commands to lower cortical centers and movement then takes place. Feedback enables units within these hierarchies to communicate, and therefore movement can be controlled and adjusted. A starting point for looking at models of control is open- and closed-loop modes of control.

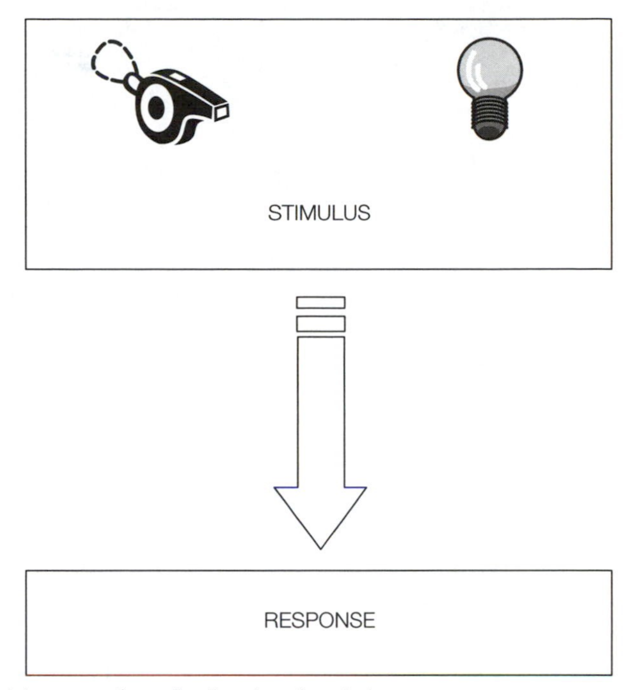

Fig. 2. Model representing reflex theories of control.

Open- and closed-loop modes of control

Adams (1971) proposed two modes of control: open and closed loop. In a closed-loop control system (*Fig. 3*) the control center issues an initial command to the effectors that is sufficient to initiate the movement. The execution and completion of such movements are however dependent upon feedback as this controls the ongoing movement. In an open-loop system (*Fig. 4*) feedback is not used in the control of the ongoing movement; the movements are believed to be pre-programed.

Although more traditional explanations of goal-directed movements propose both open-loop and closed-loop components of movements, more recent theories predict less involvement of feedback (Meyer *et al.*, 1988) and others claim it plays virtually no role in the control of rapid movements (Plamondon, 1995a, b). Plamondon (1995a) suggested that 'sensory feedback is not used to control the trajectory' of aiming movements or reaching movements and that 'feedforward

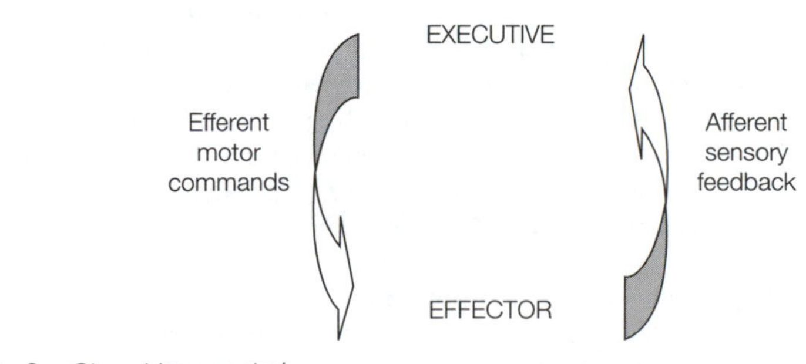

Fig. 3. Closed-loop control.

Fig. 4. Open-loop control.

control emerges through practice and learning' (p.216). However, several recent studies still support the two-component model of limb control (Helsen *et al.* 1998a, 2000; Starkes *et al.*, 2002), and this pattern of results has been found regardless of whether one performs aiming with the right or left hand (Helsen *et al.*, 1998b), whether target eccentricity is varied from 10 cm to 40 cm (Helsen *et al.*, 1997) and even when movement complexity is altered (Helsen *et al.*, 1998b).

The speed at which the movement has to be performed has a major influence on the mode of control employed. The most basic distinction in motor control is that between ballistic and controlled movements. Ballistic movements are rapid movements that do not allow for correction during the movement and therefore need to be preprogramed. Feedback is not used during a ballistic movement, but should the same movement be repeated the performer would make changes as a result of the first trial. Controlled movements are relatively slower and changes are made during the movement. Over one century ago, Woodworth (1899) first explicitly distinguished between the two phases in voluntary goal-directed aiming movement and was the first to compare fast and slow movements. Woodworth (1899) asked subjects to move a pencil back and forth through a slit with the direction reversing once two marks had been passed. The rate of the moving pencil was controlled by a metronome. Subjects performed one set with their eyes open and another set with their eyes closed. He discovered that when subjects had their eyes open, mean absolute error decreased as velocity decreased. Thus, accuracy improved as movements slowed in the eyes-open condition but not in the eyes-closed condition. Woodworth concluded that in the eyes-closed condition subjects' movements were entirely preprogramed; he referred to this as an initial impulse. In the eyes-open condition he concluded that movements were both preprogramed and corrected with current control (visual feedback).

Generalized motor program

Taking a hierarchical stance, movements are stored in memory as plans or motor programs for movement. It was Keele (1968, p.387) who defined a motor program as 'a set of muscle commands that are structured before a movement sequence begins'. However, the idea of a motor program being stored for every movement obviously has storage problems. The work of Keele stimulated a mass of research in this area and a broadening of the definition that took into account sensory feedback and was better able to account for how we were able to store movement memory. The work of Schmidt led to the term generalized motor program (GMP), which indicated that we stored families of movements rather than individual programs. It must be remembered that theories of learning and theories of control are closely related and a full review of the generalized motor program can be found in Section H. The generalized motor program is more adaptable than the motor program and it better explains the control of movement in a more varied and dynamic environment. It also explains how we are able to produce a wide variety of similar movements without the issues of storage. It also better explains motor equivalence and how we are able to produce novel movements.

The next theories of control to be reviewed approach control from a very different stance to previous theories. Both dynamical and ecological theories take into account the role of the environment and its interaction with the individual. These theories are better able to address Sheridan's aspects of control and they provide a more flexible overview of control that considers self-organization of the movement system. Dynamical and ecological theories are better able to explain how we control complex movement in an adaptive and demanding environment regardless of our level of expertise or experience.

Dynamical systems theories

Dynamical systems theories of control approach control in terms of a problem; that is how multiple degrees of freedom are controlled. How a number of units (joints, muscles) are controlled and co-coordinated to produce skilled movement needs further explanation. Bernstein (1967) raised a number of questions about the control and coordination of movements. He tried to characterize the study of movements in terms of the problems of co-coordinating and controlling a complex system of biokinematic links. In particular he worked on the issue of degrees of freedom. In an arm movement there are many structures that could be controlled at any one time. These could be the joints, muscles, alpha–gamma linkages or motor units. If we just consider the joints of the arm there are three degrees of freedom of movement at the shoulder joint, two degrees of freedom of movement at the elbow, and two degrees of freedom of movement at the wrist, this can be seen in *Fig. 5*.

Kelso (1982) states that if you take into account motor units per muscle (100) and joints in the arm, there are around 2600 degrees of freedom to be regulated at any one time. If these units are being controlled separately this would be a lot for any controlling mechanism. Kelso (1982) uses the example of a system with two elements (*Fig. 6*) operating in two-dimensional space. As there are two axes each element has a value on each axis, so element A has a y value and an x value, as does element B. As a result in (a) the two elements create a system that has four degrees of freedom of movement. By linking the two together (b), the positions of A and B are now dependent on the position of each other and the degrees of freedom of movement is reduced to three.

However, movement control is not that simple; movement takes place in a variety of different contexts and changing conditions. Bernstein (1967) identified three main sources of context-conditioned variability. Firstly, variability due to anatomical factors; secondly, mechanical variability; and thirdly, physiological variability. Bernstein considered this to be how many degrees of freedom can be

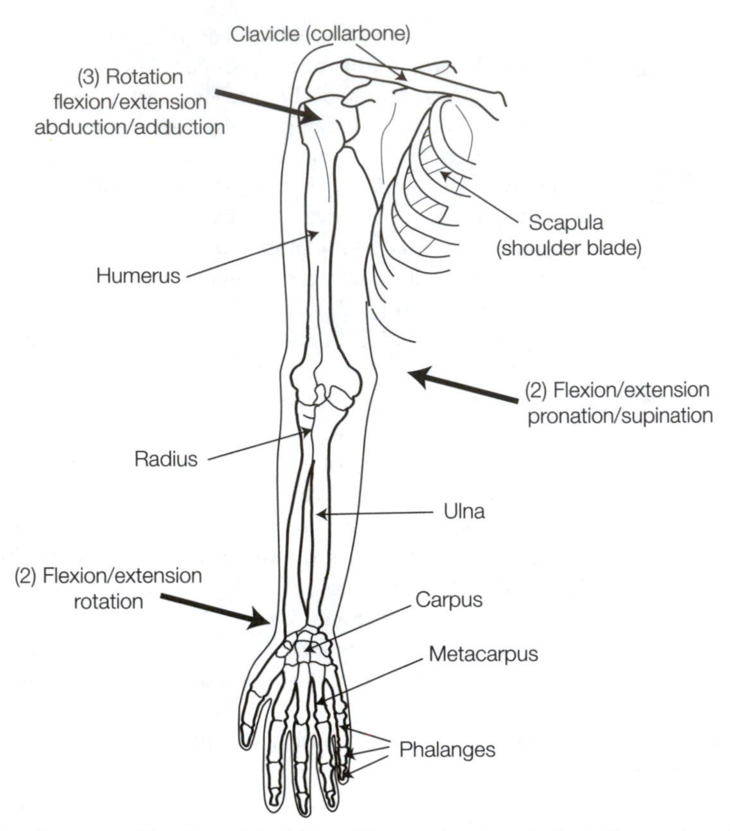

Fig. 5. Degrees of freedom of the joints of the arm (numbers indicate the number of degrees of freedom).

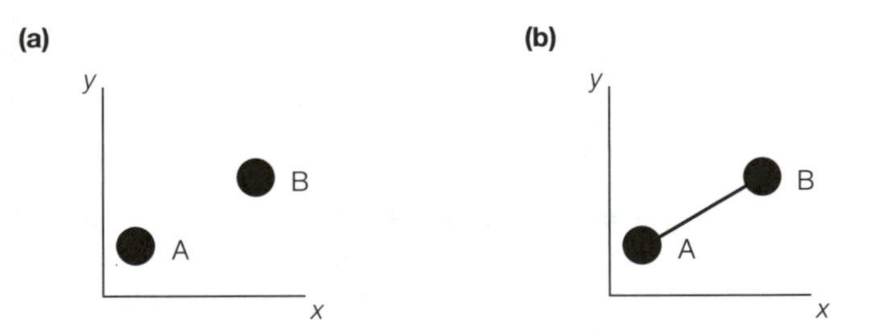

Fig. 6. How linking two elements (a) can reduce the degrees of freedom of movement. (b) demonstrates that by linking elements together control is potentially simpler as fewer degrees of freedom of movement have to be controlled at any one time. Adapted from Kelso (1982).

regulated systematically in varying control contexts without totally overloading the system. He suggested that a given nervous system program will produce different outcomes in different situations. He referred to a 'motor image', a central representation of the movement by the nervous system. The 'motor image' represented the form of movement to be achieved not the impulses needed to achieve it. He believed that proprioception played a vital part as it contributed to the representation of the movement. He also believed that one way of controlling a large number of degrees of freedom of movement was to organize actions into synergies (*Fig. 7*).

Turvey (1977) proposed the concept of coordinative structures and this concept helps our understanding of how we control a number of movement variables in varying movement contexts. A coordinative structure is a group of muscles spanning several joints that is constrained to act as a single functional unit. Coordinative structures can be described as groups of muscles, often traversing several joints, which are constrained by the intrinsic structure of the mechanical system to act as a coherent unity (Davids *et al.*, 1994). The formation of coordinative structures assists us with controlling complex movements. Tuller *et al.* (1982) give numerous examples to illustrate this point. They give the example of steering a car where the front wheels are linked together. If the two wheels had to be controlled separately this would cause major problems for the driver. Analysis of the coordinative structures in animals such as the gait of a horse where changes of speed happen instantly promote the argument that there are biomechanical as well as cortical decisions supporting these changes. Other studies (Kugler and Turvey, 1987; Kelso *et al.*, 1991) on rhythmic hand and finger movements have shown how human movement systems can be described through the principle of self-organization.

The concept of coordinative structures can help overcome the problem of controlling innumerable degrees of freedom. However, another movement issue that has to be considered is: how do we control a particular action in different contexts? The coordinative structure must be adjustable. Tuller *et al.* (1982) among others have argued that a coordinative structure is a device that adjusts itself automatically to changing external conditions and achieves the same result. The ability to do this has been compared to a simple mass spring

Fig. 7. Cycling involves the control and coordination of multiple degrees of freedom in a changing environment.

model, which is a spring attached at one end to a fixed support and at the other end to a mass. Whether stretched (pulled) or compressed (pressed), the spring calibrates at a constant length regardless of the initial conditions and comes to rest back at it original starting position. This oscillatory system does not need any controller directing it. There is some evidence that some movements of the biological system exhibit the same properties as oscillatory systems. Movements at the elbow (Feldman, 1966) and finger movements (Kelso, 1977) act in an analogous way to a mass spring system. Kelso (1977) conducted an experiment which tested this theory. Subjects moved their forefinger from side to side over a distance that they selected. Their finger was then passively moved by the experimenter to a different starting position and the subject was asked to produce either the amplitude or the final position of the first movement. Kelso discovered that subjects could return to the same final position of the first movement with a high level of accuracy, but could not produce the same amplitude very accurately. Subjects could also return to the same final position of the first movement without being able to see or feel their finger's location. This research supports the mass-spring theory of control and explains how movement can be adjusted automatically depending on changing initial conditions.

The above theories help our understanding of how we control a number of degrees of freedom of movement, in a number of varying contexts, influenced by many variables. A coordinative structure explains how we control the number of degrees of freedom. The analogy of the mass spring is used to describe how a biological system can be self-regulating.

Ecological theories

The ecological perspective on motor control is based on the fact that the perceptual and motor systems evolved in natural environments, and that a good starting point is the physical task to be performed. An ecological approach to the control of actions considers how information available within the environment supports the coordination of movements.

As we move in the environment, the pattern of light on the retina varies depending on the layout of the environment, the level of illumination and the way in which we are moving. Gibson (1966, 1979) argued that the layout of the environment is specified by the optical array that it omits, that is the light that is perceived by an individual. The color, texture, shape and angle of inclination will affect the light pattern, or optical array in a characteristic way. This provides the performer with the perceptual information that they need to control movements in differing environments and contexts. Information from the environment is used by the performer to regulate motor activity. If a person moves forward in a stable environment, then the optical array will undergo an outward flow pattern. When stationary a person who then rotates the body will experience a local expansion of the optical array. Thus there is an invariant pattern of change for every type of movement. This information is vital to the performer and plays a significant role in the control of actions. Lee and Aronson (1974) and Lee and Lishman (1975) tested Gibson's theory. They constructed a room with a stable floor surrounded by walls which could be moved forwards and backwards. They wanted to see how the subjects reacted when the optical array was manipulated by moving the walls slowly while the subject stood still. They found that subjects swayed and that the direction of the optical illusion influenced the direction of the sway. When the walls moved forward they swayed backwards; as the wall approached the subject the retinal image size increased as it would if you were walking towards the wall; as a result the subject sways backwards. Further work with children showed that they use visual information in the same way as adults. However, as they do not have the same proprioceptive experience as adults, the effect of moving the room was much more dramatic. When the room was moved forwards or backwards some children actually fell over in the opposite direction or swayed and had to take a step to steady themselves.

Lee and coworkers concluded, as did Gibson, that kinesthesis is not totally directed from within the body (proprioception) but can be controlled by vision (visual kinesthesis). Lee expanded this work in 1976 by calculating a precise relationship between the optical flow and movement in the external environment. He discovered that when an object is approaching the eye, the greater the speed of approach, the higher the rate of retinal image expansion (Lee, 1976). The higher the retinal image expansion rate then the less time there is until contact. Therefore time to contact can be calculated by the visuomotor system and the appropriate response made. Lee named this formula *tau* 'The inverse of the rate of expansion of the retinal image of an object equals the time remaining until contact is made with the object' (Rosenbaum, 1991).

According to Gibson (1979), a lawful account of perceiving and acting in dynamic contexts should be based upon a theory of a mutually constraining relationship between animals and their environments. Gibson (1979) theorized that a lawful relationship exists between the properties of the environment and the structure of surrounding energy distributions. The layout of the environment imposes a rich spatio-temporal order upon the surrounding energy flows (*Fig. 8*), particularly the crucial source-optical energy. Gibson (1966, 1979) argued that the layout of the environment is specified by the optical array that it

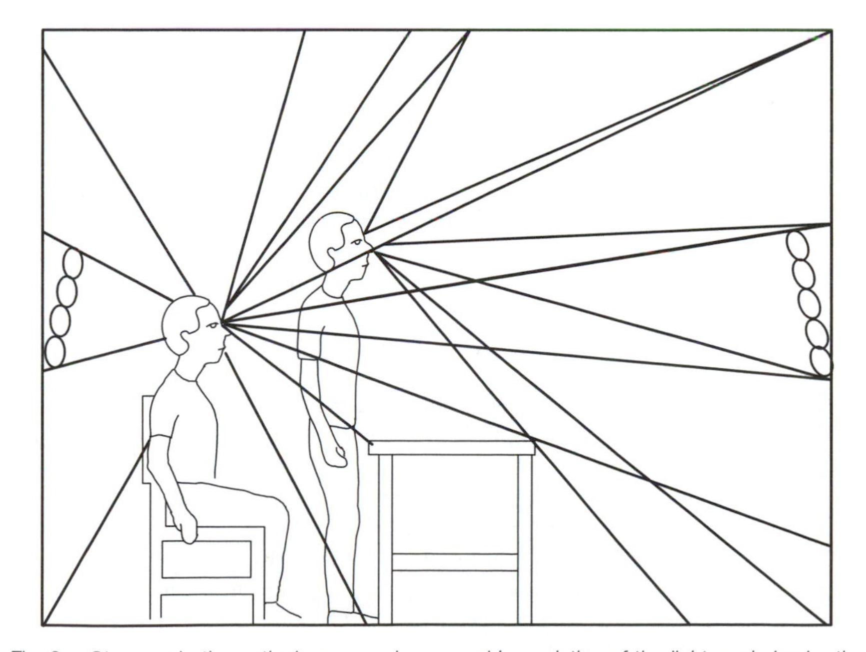

Fig. 8. Changes in the optical array can be caused by variation of the light, and also by the movement of an individual.

emits, that is, the light that is perceived by an individual, providing the performer with the perceptual information they need to control movements in differing contexts. Gibsonian theorists have highlighted the relationship between perception and action as direct and cyclical, in that information from the environment is used to regulate motor activity.

- *Flow field*: includes perceptual information from vision, haptic, proprioception, and auditory systems.
- *Force field*: refers to how we effect change in the environment through action.

The ecological approach to control has two main ideas that distinguish it from other theories. One is the concept of affordances, the other is the concept of invariants or higher-order properties of surrounding energy flow. Gibson (1966, 1979) introduced the idea of affordances. He stated that objects invite a specific action within a specific context. Therefore objects afford certain actions, and a cube, for example, is seen by the performer not as a cube but as an object that affords differing types of manipulation.

> The object attributes are represented therein as affordances, that is, to the extent that they trigger specific motor patterns for the hand to achieve the proper grasp. In addition, this function does not seem to imply binding of object attributes into a single entity. Instead, each attribute contributes to the motor configuration of the hand by selecting the relevant degrees of freedom. (Jeannerod, 1994, p.8)

Affordances represent possibilities in different contexts within the scope of the performers' individual functional ability. This is a natural way in which the multiple degrees of freedom can be controlled. Kelso and Tuller (1981) also believe that situational conditions can provide constraints. An object can only 'afford' so much and will have different constraints in differing contexts and vary for each individual. Affordances are invitations to act within specific contexts (Gibson, 1979). For example, the athlete in the 3000 m steeple chase may see a barrier as an object that affords hurdling; another athlete may see it as an object that affords placing a foot on top and pushing off. The style adopted or afforded depends on the interaction between the expertise of the athlete, the task and the context. In *Fig. 9* the task influences the grasp used by the performer for example.

Affordances for perception and action

The concept of affordances is the central aspect of the ecological approach. In his ecological approach to perception, Gibson (1979) proposes that from the origins of development what are perceived are the affordances of the environment: '. . . what it offers the animal, what it provides or furnishes, for ill or for good' (p.127). Hence, an affordance 'refers to the fit between an animal's capabilities and the environmental supports and opportunities (both good and bad) that make possible a given activity' (Gibson and Pick, 2000, p.15). Affordances are therefore possibilities for action; any given task, in a particular context, can be achieved in a number of ways depending upon the individual.

Fig. 9. Performing manual tasks requires control of the hands, a high level of coordination between other limbs and postural control. Different tools and shapes afford a different grasp and the grasp employed depends upon the task.

Gibson's theory of affordances therefore puts great importance on the link between perception and action. It is based on the premise that perceptual information is primarily constrained by the meaning of action (Rochat, 1995). If an animal guides its movement by affordances, it must be capable of perceiving the relationship between environmental properties and the properties of its own action system. One of the implications of this assumption is that actions are 'body-scaled' and perceptions are driven by intrinsic body measures as opposed to extrinsic or absolute measures. For example, Warren (1984) found that when a person was confronted with the problem of climbing the stairs, the 'climbability' was specified by a critical or boundary ratio between the person's leg length and the tread height. Ratios smaller than the critical ratio specify climbability, ratios larger than the critical height specify non-climbability. Hence body-scaled ratios can be used as a critical determinant of action choice, a change beyond the critical ratio determining a new class of action. Critical body-scaled ratios have been found for a variety of action patterns such as gait (Alexander, 1994), sitting height (Mark, 1987), walking through apertures (Warren and Whang, 1987) and reaching (Carello *et al.*, 1989). Thus, critical ratios can describe when a shift in behavior occurs (Van der Kamp *et al.*, 1998). In contrast, the switching between movement patterns may also emerge from the changes in the constraints imposed upon the action.

To summarize so far, affordances for action in an environment are specified by optical invariants and are contained within a unique frame of reference for each individual. Descriptions of the state of the environment can only ever be 'frame dependent' because affordances are perceived in relation to relevant properties of an individual (Turvey, 1986). Therefore, a mutually constraining relationship exists between intentionality, perception and action within an ecological frame of reference for each other.

Perception–action coupling

The environment provides opportunities and resources for action, and information to guide subsequent action. The natural physical approach to motor control suggests that perceptual information coordinates the spatio-temporal characteristics of the athlete's behavior in achieving an environmental goal. The specific movement of the performer creates perceptual information that is used to 'guide' the evolving dynamics of the neuromuscular system to the most adaptable pattern required for successful task execution. Gibson has considered the role of information in the control of movements to objects of varying location and orientation. Visual information is gained from the optic array, which has varying intensities dependent upon orientation and direction and the object from which it emits. This provides visual information, which assists control. Lee *et al.* (1982) used an ecological framework for under-

Fig. 10. In long jumping an accurate take-off is vital.

standing timing behavior in elite long jumpers. They argued that elite jumpers subconsciously used light energy from the take-off board (flow field) to modulate their stride pattern (force field) for take-off (*Fig. 10*). Large irregularities of the stride pattern in the last few strides are seen as evidence of using visual regulation.

An ecological perspective on the control of catching assigns primary importance to the fact that, for successful catching of a ball traveling on a relatively unpredictable trajectory, all the necessary information on which the spatial positioning and timing of the grasp of the hand are made must, by definition, be available in the optic array (Savelsbergh and Whiting, 1996). Indeed the narrow time window in which the performer has to initiate the grasp for a successful catch dictates that the performer must have access to predictive temporal information. Therefore there is a tight coupling between the action and invariant parameters of the environment.

Coordinative structure hypothesis

The natural physical approach to perception and action views the human biokinematic system as 'self-reading' and 'self-writing' (Kugler, 1986). There is no need for instructional devices for prescribing the pattern of contractions to specific muscles due to feedback and feedforward loops. Alternatively what is suggested is that coordinative structures are formed representing attractor states which are changing task solutions and do not need to be represented within the system.

Coordinative structures allow the performer to take advantage of the intrinsic design principle of the biomechanical system by temporarily assembling 'task-specific' devices in goal-directed behavior. These task-specific devices can be viewed as connections between component parts of the motor system. However, as a human performer acts in a dynamic environment, the energy that enters the system is ever-changing, and the temporal symmetry to which the system has been attracted becomes unstable. A modified version of the original action is therefore assembled. Information from the environment allows the system to move onto an appropriate attractor in order for a successful action to the problem or task to be found. In human performance, the information that is used to switch between attractor points surrounds the athlete.

In an attempt to assess the strength of upper-limb synchronization, researchers have explored the capability of human subjects to make different but simultaneous movements in limbs. Kelso *et al.* (1983) examined what would happen when an individual was asked to move each hand to a target that had a different index of difficulty. The tendency was for the two hands to come together regardless of the fact that they were moving to different-sized targets or targets that were differing distances away. In an earlier study, Kelso *et al.* (1979) found that limbs interacted by rescaling the temporal structure to obtain a common timing basis. Kelso *et al.* (1979) investigated bimanual aiming movements over different distances (6 cm and 24 cm). In the first of three experiments reported, subjects were required to move their index fingers from a centrally located position so the lateral targets could be touched as quickly and as accurately as possible. Subjects initiated two-handed movements almost simultaneously, the largest interlimb difference in reaction time (RT) being just 8 ms. Within-subject correlations for RTs between right and left hands in the two-handed conditions ranged from 0.95 to 0.97. Moreover, movement time (MT) data also showed that the subject's hands moved at different speeds to different target endpoints.

Activity

Draw the diagram below on a piece of paper

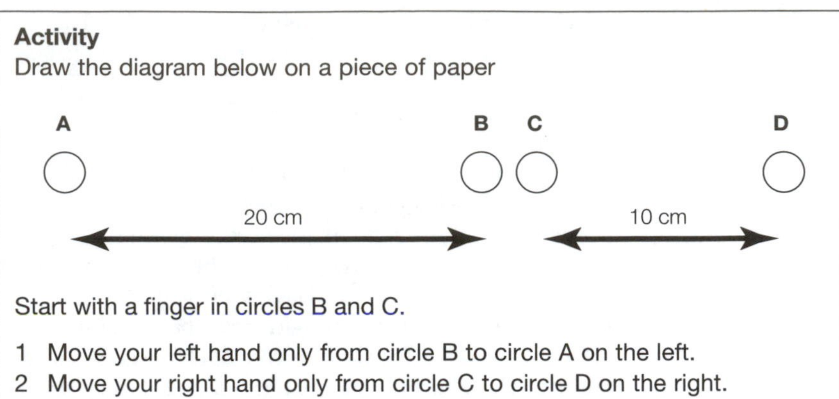

Start with a finger in circles B and C.

1 Move your left hand only from circle B to circle A on the left.
2 Move your right hand only from circle C to circle D on the right.
3 Now start with your fingers in B and C and move your left hand to circle A and back and your right hand to circle D and back.

You should find that when moving both hands together you start and finish at the same time as in the Kelso *et al*. (1983) experiment.

Thus one of the main findings of the study was that even when each limb was required to move different distances to the targets, time to peak velocity (TPV) and time to peak acceleration (TPA) for each limb showed a synchronous pattern, despite the fact that both limbs traveled at entirely different speeds. These findings led Kelso *et al*. (1979) to conclude that the brain sends signals to muscle groupings superimposed across the two limbs. Timing was seen as an essential variable (i.e. a variable that reflects the overall behavioral organization during action) in the control of bimanual coordination.

In particular the work of Kelso *et al*. (1979) has been interpreted as support for the theoretical notion that actions are controlled via coordinative structures (Turvey, 1977). Thus, during bimanual movements, rather than being controlled separately, the limbs are controlled via across-limb functional groups of muscles that form significant units of action thus reducing the cognitive load placed upon the executive system (Kelso *et al*., 1979). Nevertheless, some studies have shown a partial distribution of this common temporal structure with increasing spatio-temporal divergence of bimanual movements, and thus other explanations have been investigated.

Conclusion

Theories of motor control provide the foundation on which practice should be based. These theories are not abstract ideas that have no relevance to the applied context. As we have seen in this section there are large differences between reflex, hierarchical, dynamical, and ecological theories. There are still conflicting opinions as to how relevant the general motor program is, and as to how appropriate dynamical and ecological theories of control are. Further research will no doubt continue to explore the pro and cons of each in a variety of movement contexts.

Further reading

Bonfiglioli, C., Pavani, F. and Castiello, U. (2004) Differential effects of cast shadows on perception and action. *Perception* **33**, 1291–1304.

Davids, K., Handford, C. and Williams, M. (1994) The natural physical alternative to cognitive theories of motor behavior: an invitation for interdisciplinary research in sports science? *J Sports Sci* **12**, 495–528.

Piet, C.W. and van Wieringen, C.W. (1996) Ecological and dynamical approaches to rehabilitation. *Hum Mov Stud* **15**, 315–323.

References

Adams, J.A. (1971) A closed loop theory of motor learning. *J Mot Behav* **3**, 111–149.

Alexander, R.M. (1994) Walking and running. *Sci Amer* **72**, 348–354.

Bernstein, N. (1967) *Coordination and Regulation of Movements*. Pergamon Press, New York.

Carello, C., Grosofsky, A., Reichel, F.D., Solomon, H.Y. and Turvey, M.T. (1989) Visually perceiving what is reachable. *Ecol Psychol* **1**, 27–54.

Davids, K., Hanford, C. and Williams, A.M. (1994) The natural physical alternative to cognitive theories of motor behavior: an invitation for interdisciplinary research in sports science? *J Sports Sci* **12**, 495–528.

Feldman, A.G. (1966) Functional tuning of the nervous system during control of movements or maintenance of a steady posture-III: mechanographical analysis of the execution of man of the simplest motor tasks. *Biophysics* **11**, 766–775.

Gibson, J.J. (1966) *The Senses Considered as Perceptual Systems*. Houghton-Mifflin, Boston.

Gibson, J.J. (1979) *The Ecological Approach to Visual Perception*. Lawrence Erlbaum Associates, Hillsdale, NJ.

Gibson, E.J. and Pick, A.D. (2000) *An Ecological Approach to Perceptual Learning*. Oxford University Press, Oxford.

Helsen, W., Starkes, J.L. and Beukers, M.J. (1997) Effects of target eccentricity of temporal costs of point of gaze and hand aiming. *Motor Control* **1**, 161–177.

Helsen, W., Starkes, J.L., Elliot, D., and Beukers, M. (1998b) Manual asymmetries and saccadic eye movements in right-handers during single and reciprocal aiming movements. *Cortex* **34**, 513–529.

Helsen, W., Starkes, J.L., Elliot, D. and Ricker, K. (1998a) Temporal and spatial coupling of point of gaze and hand coordination in aiming. *J Mot Behav* **30**, 249–259.

Helsen, W., Starkes, J.L., Elliot, D. and Ricker, K. (2000) Coupling of eye, finger, elbow and shoulder movements during manual aiming. *J Mot Behav* **32**, 241–248.

Jeannerod, M. (1994) Object oriented action. In: Bennett, K.M.B. and Castiello, U. (eds) *Insights into the Reach to Grasp Movement*. North Holland, Amsterdam, pp.3–19.

Keele, S.W. (1968) Movement control in skilled motor performance. *Psychol Bull* **70**, 387–403.

Kelso, J.A.S. (1977) Motor control mechanisms underlying human movement reproduction. *J Exp Psychol* **3**, 529–543.

Kelso, J.A.S. (1982) *Human Motor Behavior*. Erlbaum, Hillsdale, NJ.

Kelso, J.A.S., Buchanan, J.J. and Wallace, S.A. (1991) Order parameters for the neural organization of single, multijoint limb movement patterns. *Exp Brain Res* **85**, 432–444.

Kelso, J.A.S., Southard, D.L. and Goodman, D.L. (1979) On the co-ordination of two handed movements. *J Exp Psychol Hum Percept Perform* **5**, 229–238.

Kelso, J.A.S., Southard, D. and Goodman, D. (1983) On the nature of human interlimb coordination. *Science* **203**, 1029–1031.

Kelso, J.A.S. and Tuller, B.H. (1981) Towards a theory of apractic syndromes. *Brain Lang* **12**, 224–245.

Kugler, P.N. (1986) A morphological perspective on the origin and evolution of movement patterns. In: Wade, M. and Whiting, H.T.A. (eds) *Motor Development in Children: aspects of coordination and control*. Martinus Nijhoff, Dordrecht, pp.459–527.

Kugler, P.N. and Turvey, M.T. (1987) *Information, Natural Law and the Self-assembly of Rhythmic Movement*. Erlbaum, Hillsdale, NJ.

Lee, D.N. (1976). A theory of visual control of braking based on information about time to collision. *Perception* **5**, 437–459.

Lee, D.N. and Aronson, E. (1974) Visual proprioceptive control of standing in human infants. *Percept Psychophys* **15**, 529–532.

Lee, D.N. and Lishman, J.R. (1975) Visual proprioceptive control of stance. *J Hum Mov Stud* **1**, 87–95.

Lee, D.N., Lishman, J.R. and Thomson J.A. (1982) Regulation of gait in long jumping. *J Exp Psychol Hum Percept Perform* **8**, 448–459.

Mark, L.S. (1987) Eye height-scaled information about affordances: a study of sitting and stair climbing. *J Exp Psychol Hum Percept Perform* **13**, 361–370.

Meyer, D.E., Abrams, R.A., Kornblum, S., Wright, C.E. and Smith, J.E.K. (1988) Optimality in human motor performance: ideal control of rapid aimed movements. *Psychol Rev* **95**, 340–370.

Plamondon, R. (1995a) A kinematic theory of rapid human movements: 1. Movement representation and generation. *Biol Cybern* **72**, 295–307.

Plamondon, R. (1995b) A kinematic theory of rapid human movements: 2. Movement representation and generation. *Biol Cybern* **72**, 309–320.

Rochat, P. (1995) Perceived reachability for self and others by 3–5 year old children and adults. *J Exp Child Psychol* **59**, 317–333.

Rosenbaum, D.A. (1991) *Human Motor Control*. Academic Press, San Diego.

Savelsbergh, G.J.P. and Whiting, H.T.A. (1996) Catching: a motor learning and developmental perspective. In: Heuer, H. and Keele, S. (eds) *Handbook of Perception and Action: motor skills*. Academic Press Limited, London, pp.461–497.

Sheridan, M.R. (1984) Planning and controlling simple movements. In: Smyth, M.M. and Win, A.M. (eds) *The Psychology of Human Movement*. Academic Press, San Diego, pp.47–82.

Sherrington, C. (1906) *The Integrative Nature of the Nervous System*. Yale University Press, New Haven, CT.

Starkes, J., Helsen, W. and Elliott, D. (2002) A ménage à trois: the eye, the hand and on-line processing. *J Sports Sci* **20**, 217–224.

Tuller, B., Turvey, M.T. and Fitch, H.L. (1982) The Bernstein perspective: II. The concept of muscle linkage or coordinated structure. In: Kelso, J.A.S. (ed) *Human Motor Behavior*. Erlbaum, Hillsdale, NJ, pp.253–281.

Turvey, M.T. (1977) Preliminaries to a theory of action with reference to vision. In: Shaw, R. and Bransford, J. (eds) *Perceiving, Acting and Knowing: towards an ecological psychology*. Erlbaum, Hillsdale, NJ, pp.211–265.

Turvey, M.T. (1986) Intentionality: a problem of multiple reference frames, specificational information, and extraordinary boundary conditions on natural law. *Behav Brain Sci* **9**, 153–155.

Van der Kamp, J., Savelsbergh, G. P. and Davis, W.E. (1998) Body-scaled ratio as a control parameter for prehension in 5- to 9-year-old children. *Dev Psychobiol* **33**, 351–361.

Warren, W.H. (1984) Perceiving affordances: visual guidance of stair climbing. *J Exp Psychol Hum Percept Perform* **10**, 683–703.

Warren, W.H. and Whang, S. (1987) Visual guidance of walking through apertures: body-scaled information for affordances. *J Exp Psychol Hum Percept Perform* **13**, 371–383.

Woodworth, R.S. (1899) The accuracy of voluntary movements. *Psychol Rev Monograph Supplement* **3**.

Section E

INFORMATION PROCESSING

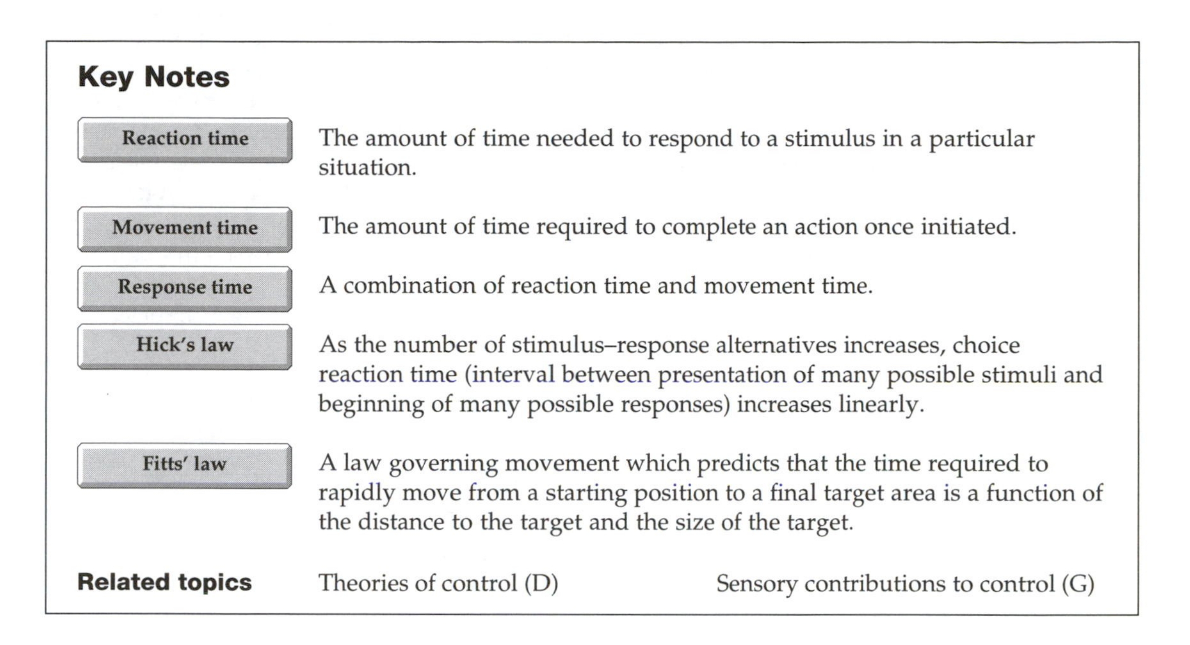

Key Notes

Reaction time	The amount of time needed to respond to a stimulus in a particular situation.
Movement time	The amount of time required to complete an action once initiated.
Response time	A combination of reaction time and movement time.
Hick's law	As the number of stimulus–response alternatives increases, choice reaction time (interval between presentation of many possible stimuli and beginning of many possible responses) increases linearly.
Fitts' law	A law governing movement which predicts that the time required to rapidly move from a starting position to a final target area is a function of the distance to the target and the size of the target.

Related topics Theories of control (D) Sensory contributions to control (G)

In many sporting situations, skilled performance is the result of correctly analysing information in the environment to determine which movement is the most appropriate.

The decisions we make are reflected in correct or incorrect movements and contribute to our ability to perform skilled movements. It is therefore logical that to do their job well physical educators, coaches and therapists have to understand the numerous and complex problems associated with motor performance. They need to understand the capacities and limitations of the performer or patient, and to understand how complex skills are learned they must appreciate how information is processed and behavior regulated. The information-processing approach provides a conceptual framework for understanding how information is perceived, established and used with respect to our short- and long-term memory, and then processed and used to produce movement. Understanding this process and its limitations also gives rise to some important principles around creating an appropriate learning environment for the child, athlete and patient.

Basic concepts

The information-processing approach describes behavior as predominantly a hierarchical process, and researchers often use the analogy of a computer to describe the distinct stages that motor control is broken down into (Stelmach, 1982). The central tenet of the information-processing approach is that when we perform an activity there are a number of mental operations performed by the learner to solve a particular problem or perform a task (Stelmach, 1982). For example, think of when you play tennis. Write down what information you think of prior to hitting the ball (e.g. pictures, sounds and feelings that convey information). You, the receiver, must perceive all that is going on around you, through using your eyes (seeing the ball travel towards you) and ears (listening for the sound of the ball on your opponent's racket strings), as well as being aware of the position of your body in space. These stimuli comprise what we call the input. Now you have all this information what are you going to do? How are you going to respond? All this information has to be analyzed and then you use your previous knowledge of such situations (memory) to make a decision about what to do based on this information (we decide and play a forehand drive). The output is the physical action we produce (a forehand drive). Internal feedback tells us about the 'feel' of the movement while we are performing it. External feedback tells us how successful we have been (did the ball land in or out) (see Section L for more information about feedback). As you can see, the information you deal with while performing this action is passed through many stages. The stages this information passes through before any movement is made can be represented by an information-processing model, a simple version of which is shown in *Fig. 1*.

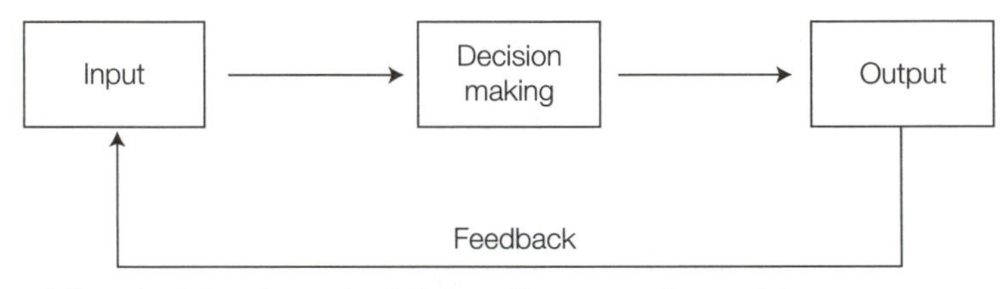

Fig. 1. A flow chart showing a simple information-processing model.

Input and stimulus identification

The brain has to decide what the information (stimulus) represents. This is referred to as the stimulus identification stage. Hence, stimulus identification is primarily a sensory (involving sense organs) stage, involving the analysis of all the various environmental information that is received from a variety of senses – stimulus detection. Imagine you are walking upstairs, something which, for many, is relatively simple. As you walk your central nervous system (CNS) is receiving information from the visual field, for example it notes if there is anybody else on the stairs, how many steps you have left until you get to the top. In addition, you receive auditory information from the sound of each step, a radio could be playing, and there is probably other background noise. With each step there is also proprioceptive input from your muscles, joints and tendons about the height of each step and where your body is in space. Despite the fact that there is an infinite array of stimuli available for the CNS to use, it is very selective, and a very sophisticated filtering process takes place determining what information is important for you to successfully walk upstairs. As we will see later, visual information plays an important role in both stimulus detection and pattern recognition. Once a stimulus has been detected we then extract a pattern or feature from the stimuli. Elements of pattern recognition are genetic and related to survival, but many are dependent on learning or experience. Once an interpretation of the stimulus has been made an appropriate response is selected (*Fig. 2*).

Decision making and response selection

The activities of the response-selection stage begin when the stimulus-identification stage provides information about the nature of the environmental stimulus. The response-selection stage has the task of deciding what movement to make,

Fig. 2. A hockey player makes a decision on how to play a shot and then responds with a particular action.

given the nature of the environment. For example, it is in this stage that the choice of movement is made from a store available, e.g. in our tennis example, should I volley the ball or go for a smash? The movement is then enacted by the effectors (muscles etc).

Output or response programing

The final stage, the response-programing stage, begins working once it has received the decision about what movement to make, as determined by the response-selection stage. The response-programing stage has the task of organizing the motor (movement) system (the muscles) for the desired movement. This involves sending brain motor commands that specify the order in which the muscles contract, with the required levels of force and timing to produce the movement effectively. If these commands are not appropriately structured then resulting movement will look uncoordinated and not fulfil the objective of the movement. Finally, the effector mechanism organizes this information and sends messages via nerve endings to the muscles that need to move (messages sent to arm muscles to take back the racket).

In light of the above, our information-processing model now looks a little like that shown in *Fig. 3*.

Reaction time and movement time

Information processing is closely linked to a person's ability to make decisions quickly (reaction time) and then transfer them into action (movement time). This is vital in many sporting situations and, indeed, many activities of daily life. The

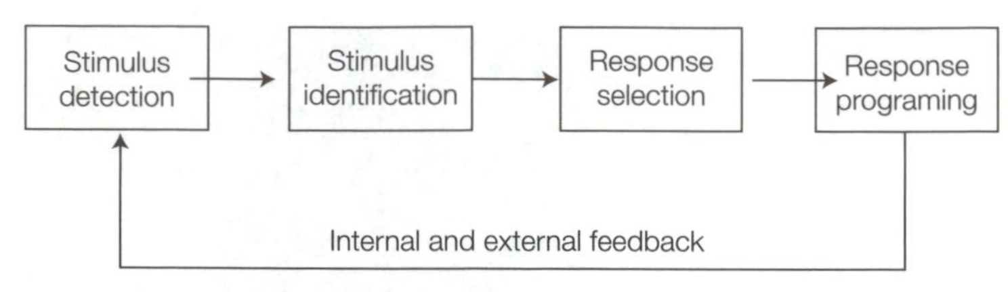

Fig. 3. A complex information-processing model.

ability to react and move quickly allows the performer to be in greater control. The use of reaction time (RT) as a measure of information processing has a long-standing tradition in human experimental psychology. A great advantage of this parameter is that it directly reflects the assumed underlying psychological/ neuronal processes. A person's reaction time is a measure of how quickly they can respond to a given stimulus, it generally reflects the time required for a stimulated receptor nerve (e.g. auditory, visual, proprioceptive) to carry the message to the brain and then for the brain to activate the appropriate muscles to initiate a response. How long it takes to react to a volleyball player spiking the ball at the net could mean the difference between a successful block of the shot or your opponents scoring a point. Similarly, how long it takes to react to a car which has braked in front of you can mean the difference between a safe stop and a collision (and an expensive insurance claim!). Once the performer or patient has made a decision about what action they are going to take in response to the stimulus, movement of one or many body parts must take place. The time taken for the movement, once initiated, to be completed is called movement time (MT). Together, response time, or total time (TT), comprises the time from the first presentation of stimuli to the movement ending, or the sum of reaction time and movement time (RT + MT = TT).

Simple, discriminative, choice reaction, and recognition reaction time

So far we have only discussed what is happening in terms of information processing when there is only one stimulus presented by itself – simple reaction time (SRT). In many sporting activities it is unlikely the performer is in a position where there is only one response option. More often than not the performer has to pick from two or more options and then react appropriately – this is called choice reaction time (CRT). In addition, there are situations when there are some stimuli that should be responded to and others that should get no response – recognition reaction time (RRT), sometimes called go/no go. This is sometimes referred to as discrimination reaction time (DRT), which is the time taken to respond when multiple signals are presented but the performer only has to respond to a specific one. For example, the coach may instruct the player to pass to a particular player even though other players are available. Another type of reaction time that is also referred to is fractionated reaction time (FRT). Here, reaction time is partitioned into premotor time and motor time and this can only be measured using electromyography (EMG).

Probably the most important reaction time study was that of Donders (1868). He was interested in the differences in timing among simple, recognition, and choice reaction time. He believed this could reveal how long a particular mental event would take to complete. Donders predicted the kinds of processes that might be involved in each task:

- *a simple time task* would require perception and motor stages – time to receive and then execute the stimulus

- *a recognition reaction time* task requires the above plus a discrimination stage
- *a choice reaction time task* requires all of the above – time to receive and execute the stimulus, and discriminate, plus a choice stage.

As expected, simple tasks take the shortest amount of time, followed by recognition tasks, with choice tasks taking the longest amount of time. Donders calculated the time required for each stage by using a subtraction technique:

- *perception and motor time*: time required for simple task
- *recognition time*: time for recognition task minus simple task
- *choice time*: time for choice task minus discrimination time.

This demonstrated a simple conclusion – more stages should require more time. In light of our information-processing model presented above (*Fig. 3*), a simple RT task, which involves the subject pressing a button when a stimulus is presented, involves two of the information-processing stages: stimulus detection and response programing. The go/no go task, in which the subject has to respond selectively to stimuli, adds stimulus identification to stimulus detection and response programing. Finally, in a choice reaction time task, the subject has to identify the stimulus from many choices and then respond accordingly, thus adding a response-selection stage (*Table 1*). However, as Friston *et al.* (1996) noted, the biggest disadvantage of this interpretation of the reaction time data is that simple, choice and recognition reaction time differ from each other by only an additional stage of processing, and that the introduction of that stage does not alter the others (*Fig. 4*).

In addition, Donders (1868) made a number of other seminal observations:

> We made the subjects respond with the right hand to the stimulus on the right side, and with the left hand to the stimulus on the left side.

Table 1. T1, T2, T3 and T4 denote the duration of the stimulus detection, identification, response selection and response programing respectively (modified from Miller and Low, 2001)

Task	Sport example	Lab example	Hypothesized stages	RT model
Simple	The start of a 100 m sprint	You respond to a light coming on by pressing the response button	1 Stimulus detection; 4 response programing	RT = T1 + T4
Choice	Passing the ball to one of many members of your team	You are seated in front of eight light bulbs, each with their own button. You must press the button corresponding to the appropriate light	1 Stimulus detection; 2 stimulus identification; 3 response selection; 4 response programing	RT = T1 + T2 + T3 + T4
Recognition	Responding to a tennis serve: ball landing in – go; ball landing out – no go	You are seated in front of a panel with five light bulbs and one response button. When the target light goes on you must press the button, but not if the four other lights come on	1 Stimulus detection; 2 stimulus identification; 4 response programing	RT = T1 + T2 + T4

Fig. 4. Reaction time can be measured using a reaction timer which can provide both reaction time and movement time data.

> When movement of the right hand was required with stimulation on the left side or the other way around, then the time lapse was longer and errors common.

This is called stimulus–response (S–R) compatibility, and this phenomenon was rediscovered nearly a century later by Fitts and Seeger (1953), and today is more commonly know as the Simon effect (Simon and Small, 1969). S–R compatibility refers to the fact that some tasks are more easy or difficult because of the particular sets of stimuli and responses that are used or because of the way in which individual stimuli and responses are paired with each other (Kornblum *et al.*, 1990). In general, the effect of the mapping between stimulus and response sets is a change in RT and/or accuracy (Lien and Proctor, 2002). S–R compatibility is one of the reasons why it is easier to catch a ball one-handed with your left hand if the ball is coming to the left-hand side of your midline.

Factors affecting reaction time

For about 120 years, the accepted figures for mean simple reaction times for college-age individuals have been about 190 ms for light stimuli and about 160 ms for sound stimuli (Brebner and Welford, 1980). However, RT is a volatile measure that is sensitive to many variables, and to fully appreciate information-processing theory, coaches and therapists have to fully understand the impact of these variables (*Fig. 5*). It is also a measure that has caused much controversy in the sporting area in terms of false starts.

Fig. 5. Game situations often place the performer in a situation where there are multiple stimuli and response alternatives.

Number of response choices

Perhaps the major variable to affect RT is the number of possible situations that arise and the number of possible responses that can be made to these situations. Essentially, if there are many possible stimuli to react to and each has many possible responses to choose from, then RT will increase. So in a one-choice situation (SRT), we would expect a shorter time to react than the two- and four-choice situations. You would expect, therefore, for a rugby player to be slower in reacting (determining which way s/he will move) when faced by many players compared to a situation when it was 1 vs. 1.

Hick (1952) found that in CRT experiments, response was proportional to log(N), where N is the number of different possible stimuli. In other words, reaction time rises with N, but once N gets large, reaction time no longer increases so much as when N was small. This relationship is called Hick's law. The full equation that describes this law is:

$$\mathrm{RT} = K \log_2 (N + 1)$$

where K is a constant (which is simple RT in most cases) and N equals the number of possible choices. This equation basically predicts that RT will increase linearly as the number of stimulus–choice alternatives increases (*Fig. 6*).

The $\log_2$ function of the equation shows that the RT increase is primarily due to the information transmitted by the possible choices, rather than the number of choice alternatives *per se*. In information theory, $\log_2$ specifies a bit of information, bit being short for a binary digit. A bit is a yes/no choice between two alternatives. Thus in a one-bit decision there are two alternatives, compared to a three-bit decision which involves eight choices. In light of this, not only does Hick's law predict that RT increases as the number of choice alternatives increase, it also predicts the specific size of increase to expect.

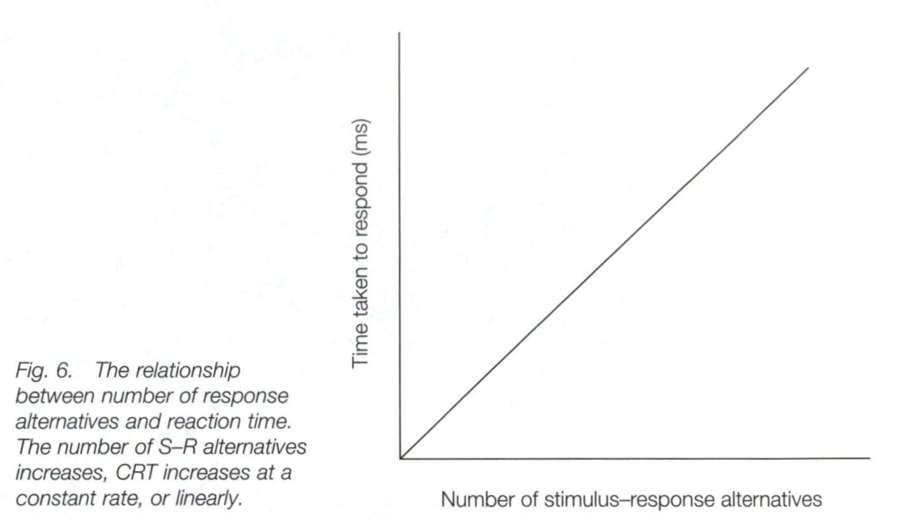

Fig. 6. The relationship between number of response alternatives and reaction time. The number of S–R alternatives increases, CRT increases at a constant rate, or linearly.

Research Highlight: Pain and Hibbs (2007)

This study looked at simple and auditory reaction time in nine athletes in four sprint start conditions:

1. normal
2. preloaded: greater force needed
3. relaxed: minimum force required in the set position
4. guess condition (anticipation).

All participants were familiar with using starting blocks and both reaction time and EMG were measured. The current International Amateur Athletic Federation criterion for a false start is a reaction time of less than 100 ms. It was interesting to note that 20% of all starts in the first two conditions were less than 100 ms. The full results with standard deviations are presented in *Table 2* below (note that the guess condition is not a true sprint start).

Table 2. Reaction times in each condition. Reproduced with permission from Pain and Hibbs (2007)

	Reaction time in each condition (ms)				
Athlete	Normal	Preloaded	Relaxed	Guess	Rejected starts
1	112 (19)	95 (22)	188 (63)	–5 (114)	2
2	97 (17)	108 (19)	104 (29)	65 (82)	2
3	122 (21)	147 (38)	199 (22)	85 (77)	1
4	119 (20)	111 (19)	122 (12)	120 (13)	0
5	106 (16)	135 (25)	158 (31)	104 (124)	3
6	113 (11)	101 (20)	123 (12)	69 (64)	2
7	107 (21)	95 (18)	150 (28)	88 (77)	3
8	82 (17)	96 (10)	95 (12)	22 (78)	7
9	94 (12)	104 (19)	123 (13)	82 (63)	2

These results indicate that athletes can produce reaction times that are less than 100 ms that are not due to anticipation.

Presentation of stimuli in rapid succession

Telford (1931) was the first to demonstrate that when people respond to each of two successive stimuli, the response to the second stimulus often becomes slower when the interval between the two stimuli is reduced, i.e. the RT for the second movement is slower than the first. Telford termed this slowing the psychological refractory period (PRP). In a typical PRP experiment, two stimuli are presented (S1, S2) which are separated by a stimulus onset asynchrony (SOA), and the person has to make a response to each of the stimuli. However, as this SOA shortens, the RT to the second stimuli becomes longer. This slowing has been observed in a great variety of tasks, including both RT (Telford, 1931) and CRT tasks (Creamer, 1963), and although most of the early experiments involved two manual responses, recent work has shown that a PRP effect can be found when the pair of tasks used require different responses (Levy *et al.*, 2006).

An example of the PRP can be seen in *Fig. 7* using the sport of tennis as an example. Looking at *Fig. 7* you will see that the first stimulus (S1) is the preparation of your opponent for a smash shot. Your reaction (R1) to this stimulus is to move towards the back of the court; however, as you begin to move, your opponent actually moves towards the ball to play a drop shot (S2). The time in between S1 and S2 is the SOA. You now have to respond to the drop shot by moving towards the net to play a return shot (R2). Essentially, this situation slows your RT, as the first piece of information needs to be cleared before the second can be processed (more comprehensive reviews of theoretical explanations of PRP are available from Pashler, 1994; Meyer and Kieras, 1997a, b; Lien and Proctor, 2002).

Other factors influencing reaction time

Even if we ignore the fact that things like S–R compatibility and the number of choices affect reaction time, there are other factors that can either shorten or lengthen a performer's reaction time. For example, our chronological age affects

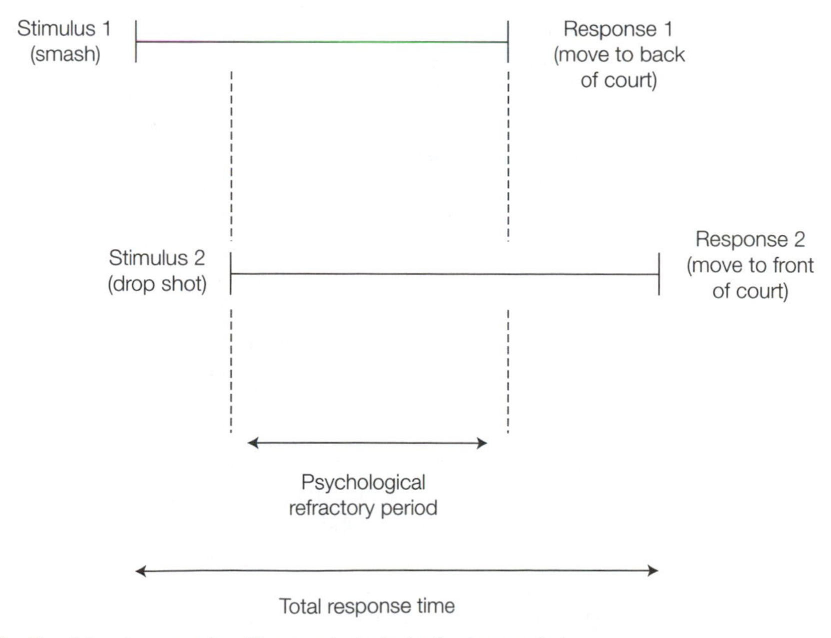

Fig. 7. A tennis example of the psychological refractory period.

RT. Simple reaction time tends to shorten from infancy into the late 20s; it then increases (and also becomes more variable), albeit slowly at first until we get to our 50s or 60s, and then increases at a more rapid rate when we are in our 70s and beyond (Canfield *et al.*, 1997; Hultsch *et al.*, 2002; Luchies *et al.*, 2002; Rose *et al.*, 2002; Der and Deary, 2006). Below is a short review of other factors that affect RT.

Handedness

It has long been suggested that the hemispheres of the cerebrum are specialized for different tasks (Sperry, 1961). For example, the left hemisphere is regarded as the verbal and logical brain, and the right hemisphere is thought to govern creativity and spatial relations (Weinstein and Graves, 2002). In addition, we know that the right hemisphere controls the left hand, and the left hemisphere controls the right hand. In light of this, it seems reasonable to hypothesize that the left hand should be faster at reaction times involving spatial relationships (such as pointing at a target).

Dane and Erzurumluoglu (2003) studied sex and handedness differences in eye–hand visual reaction times in 270 right-handed and 56 left-handed young handball players. Overall, their data showed that the left-handers had superiority over the right-handers, but there was no difference between the right eye–right hand visual reaction times of the right- and left-handers. These results are consistent with the study performed by Dane and Gumustekin (2002), who reported that the shifting distance of the left eye and the distance of focusing points of two eyes in the horizontal plane was greater in the right-handers than in the left-handers, but there is no difference between the shifting distance of the right eye of the right- and left-handers. The results of two studies suggest a left-handed advantage over right-handed in the focusing of the left eye and the left eye–left hand visual reaction times, but do not show a right-handed advantage in the focusing of the right eye and the right eye–right hand visual reaction times. The authors concluded that left-handed people have an inherent reaction time advantage.

In addition, Dane and Erzurumluoglu (2003) also found that although right-handed male handball players had faster reaction times than right-handed women, there was no such sexual difference between left-handed men and women. Recently, Silverman (2006) tested the differences in simple reaction time between males and females, and whether this has narrowed across time. A meta-analysis was conducted on 72 effect sizes derived from 21 studies (n = 15 003) published over a 73-year period. The analysis provided strong evidence for both a difference between males and females, and that the difference had narrowed over time. However, no relationship was found between the magnitude of the sex difference in RT and age or presence vs. absence of a warning signal. Silverman (2006) concluded that two factors – participation in fast-action sports and driving – are responsible for the decrease in the magnitude of the sex differences in simple visual RT across time.

Practice

Linford *et al.* (2006) examined the influence of a 6-week neuromuscular training program on the electromechanical delay and reaction time of the peroneus longus muscle in 26 men and women (the primary action of which is plantar flexion and eversion of the foot at the ankle). The participants were split into two groups. Participants in the treatment group completed a 6-week neuromuscular training program involving various therapeutic exercises. Participants in the control group were asked to continue their normal physical activity during the 6-week period.

The electromechanical delay of the peroneus longus was determined by the onset of force contribution after artificial activation, as measured by EMG and forceplate data. Reaction time was measured after a perturbation during walking. Statistical analysis showed that neuromuscular training caused a decrease in reaction during perturbed walking compared with controls. There was also a trend towards an increase in electromechanical delay, suggesting that neuromuscular training may have a beneficial effect on improving dynamic restraint during activity.

These data support those of Ando *et al.* (2004), who found that RT to a visual stimulus decreased with practice and that the effects of practice last for at least 3 weeks. Ando *et al.* (2004) examined whether the EMG RT for a key press in peripheral and central visual fields decreases with practice, and the corresponding transfer effects last for 3 weeks. Sixteen male subjects were divided into two groups, one practicing using peripheral vision, the other practicing using central vision, with each group practicing the RT tasks for 3 weeks. Data showed that the practice effects and the transfer effects were maintained as a significant decrease in RT was found over the 3-week retention interval.

Environment

Lieberman *et al.* (2006) examined the effects of a number of environmental stressors on mood and a number of cognitive performance measures, including choice reaction time. A brief, intense, laboratory-based simulation of a multistressor environment, which included sleep loss, continuous physical activity and food deprivation, was designed to simulate military operations and the stressors suffered by woodland firefighters, disaster victims and relief workers (sustained operations (SUSOPs) scenario). Data on a number of measures were collected during both the SUSOPS scenario and a control period in 13 volunteers. Results showed that there were significant decrements in choice reaction time, and overall cognitive function declined more extensively and rapidly than physical performance.

Type of sport and skill level

Baur *et al.* (2006) examined eight elite racing drivers and 10 physically active controls in a reaction and determination test requiring upper- and lower-extremity responses to visual and audio cues. Data showed racing drivers have faster reaction times than age-matched physically active controls. In contrast, Thomas *et al.* (2005) examined choice reaction times to a visual stimulus in 15 elite batsmen, 10 elite bowlers and nine controls, and found that RT was not different between cricketers and the control group.

Kida *et al.* (2005) assessed the effect of baseball experience or skill levels on simple reaction times and go/no go reaction times in 82 university students comprising 22 baseball players, 22 tennis players, and 38 non-athletes with 17 professional baseball players. Data showed that there were no differences in simple reaction time either for sports experience or for skill levels. However, when the go/no go reaction times were examined, baseball players' reaction times were significantly shorter than those of the tennis players and non-athletes. In addition, the go/no go reaction time of higher-skill baseball players was significantly shorter than that of lower-skill players, while that of the professional baseball players was the shortest.

Mouelhi Guizani *et al.* (2006) examined information processing and decision making using SRT and four-choice reaction time (4CRT) paradigms in athletes who were either experienced tennis or table tennis players, or fencers and boxers, while pedaling on a cycle ergometer at 20%, 40%, 60%, and 80% of their own maximal aerobic power (P_{max}). Statistical analysis indicated that fencers and tennis players process information faster with incrementally increasing workload, i.e. showed a faster 4CRT, while different patterns were obtained for boxers and table tennis players.

Movement time and Fitts' law

In many types of perceptual motor tasks, there is a trade-off between how fast a task can be performed and how many mistakes are made in performing the task. That is, we can either perform the task very fast with lots of errors or very slowly with very few errors. Generally, people will apply various strategies that may optimize speed, or accuracy, or combine the two. The relationship between the speed of movement and the accuracy you can maintain at any given speed is called Fitts' law, and it has been acclaimed as one of the most successful human performance models (Newell, 1990). For tasks such as aiming and pointing, Fitts' Law precisely models how time and precision affects the completion time. Thus, expressed mathematically, Fitts' law is:

$$\mathrm{MT} = a + b\,[\log_2 (2A/W)]$$

The intercept (a) and slope (b) are constants which reflect the efficiency of the system. A is its amplitude or the distance, and W is the width of the target in the direction of the movement. MT is the duration of the movement or the expected movement time, and the logarithmic term ($\log_2$) represents the accuracy of the task and is often called the index of difficulty (ID) (McCrea and Eng, 2005). In other words, Fitts' law formally models pointing speed in the form of completion time (MT) as a function of relative task precision ($2A/W$). Fitts' law mathematically explains what is known as the speed–accuracy trade-off. Essentially, the speed–accuracy trade-off is the inverse relationship between the difficulty of the movement and the speed with which it can be performed.

Anticipation

Anticipation is a strategy used by athletes to reduce the time they take to respond to a stimulus, e.g. the tennis player who anticipates the type of serve the opponent will use.

In this case the player has learnt to detect certain cues early in the serving sequence, which predicts the potential type of serve. This means the player can start to position themselves for the return earlier in the sequence than usual, and thus give themselves more time to play the shot when the ball arrives.

There are different types of anticipation. For example, Poulton (1957) distinguished between effector anticipation (knowing the position of the target or stimulus at the time the response finishes) and receptor anticipation (receptors functioning in advance of the response mechanism to determine the characteristics of the stimulus) (*Fig. 8*).

Anticipation can also be separated into spatial anticipation, (what will happen, what ball will the cricket bowler throw?) and temporal anticipation (when will it happen, when will the ball bounce, when should I start the backswing so that bat and ball meet?). However, performers must weigh cost/benefits before deciding to anticipate, what Posner (1978) calls a cost–benefit analysis. Anticipating correctly relies on experience to recognize stimuli and cues that allow the performer to process information before an event occurs; e.g. an experienced batsman would watch the bowler's hand and arm action to guess the type of delivery. A novice would watch the ball bounce before deciding which shot to play. However, if you can anticipate what kind of

Fig. 8. A Bassin timer can be used to measure anticipation.

response to give before a stimulus is presented (predicting), this can decrease reaction time, as the response-selection and/or response-programing stages can be bypassed. However, if you anticipate incorrectly, you have to stop your response and then select and program a new response; this can increase reaction time and may result in an error in performance.

Savelsbergh *et al.* (2005) conducted a study that examined anticipation in expert goal keepers. Participants had to move a joy stick in response to a penalty kick, and the number of penalties saved and frequency and time of initiation of joystick corrections were recorded. In addition, eye-tracking equipment examined visual search behaviors. Experts were more accurate in predicting height and direction of the penalty kick. They watched for longer before initiating a response and spent longer looking at the non-kicking leg (*Fig. 9*).

Box 1. Factors affecting anticipation (Jackson and Farrow, 2005)

- *Foreperiod regularity*: when the interval between warning and stimulus is constant, subjects can reduce RT significantly unless duration is too long (10–12 s). Therefore, to make anticipation unlikely use varied intervals or catch trials.
- *Foreperiod duration*: with a longer time to preview the stimulus it becomes easier to anticipate. Therefore, to make anticipation less likely the middle of a range of foreperiods is used or the duration of the foreperiod is randomized.

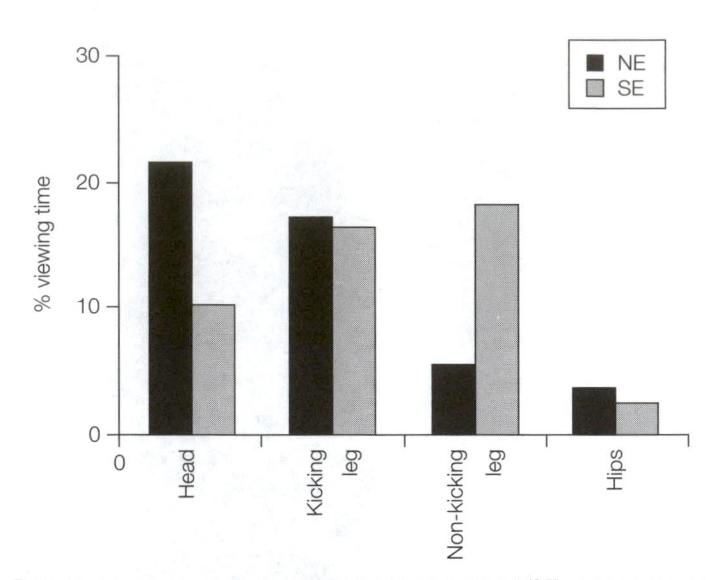

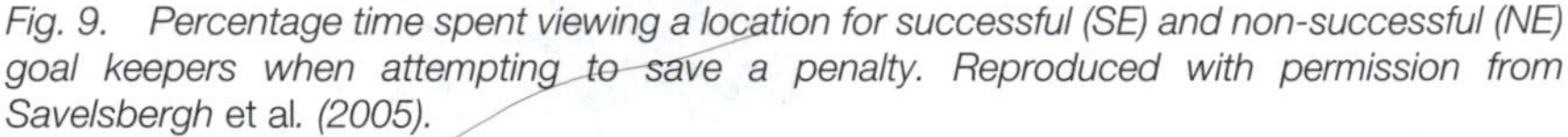

Fig. 9. Percentage time spent viewing a location for successful (SE) and non-successful (NE) goal keepers when attempting to save a penalty. Reproduced with permission from Savelsbergh et al. *(2005).*

Overall, to prevent anticipation as a performer, you should make situations unpredictable (randomization), or allow the opposition to anticipate and then do the opposite. For example, in competitive sports, opponents randomize their movements, i.e. make movements that are unpredictable, to force opponents to react rather than anticipate, or will even 'fake' certain moves so the opponent will make an incorrect prediction. In general, to prevent anticipation you should try and make the performer make use of all the information-processing stages.

Conclusion

Our ability to respond quickly and to anticipate is not only vital in the sporting arena but is also fundamental to our daily interactions with the environment. Not only do we need to respond quickly to the starter's gun, we also need to detect the direction from which a tennis ball is coming or the direction from which an ambulance is approaching. In addition, we need to be able to cross roads, get into lifts or escalators, and drive cars. All of these demand accurate responses and accurate anticipation; the cost of making an error can be disastrous. The teacher, coach, parent, and clinician must assist the learner in extracting information from the environment that enables the ability to respond and anticipate to be maximized. The coach must also teach the performer how to fool their opponents in order to ensure that their tactics are not detected and to give them the advantage.

Further reading

Masanobu, A. and Choshi, K. (2006) Contingent muscular tension during a choice reaction task. *Percept Mot Skills* **102**, 736–747.

Pain, M.T. and Hibbs, A. (2007) Sprint starts and the minimum auditory reaction time. *J Sports Sci* **25**, 79–86.

Poulter, D.R., Jackson, R.C., Wann, J.P. and Berry, D.C. (2005) The effect of learning condition on perceptual anticipation, awareness, and visual search. *Hum Mov Sci* **24**, 345–361.

Terao Y., Okano T., Furubayashi, T. and Ugawa, Y. (2006) Effects of thirty-minute mobile phone use on visuo-motor reaction time. *Clin Neurophysiol* **117**, 2504–2511.

References

Ando, S., Kida, N. and Oda, S. (2004) Retention of practice effects on simple reaction time and central visual fields. *Percept Mot Skills* **98**, 897–900.

Baur, H., Müller, S., Hirschmüller, A., Huber, G. and Mayer, F. (2006) Reactivity, stability, and strength performance capacity in motor sports. *Br J Sports Med* **40**, 906–910.

Brebner, J.T. and Welford, A.T. (1980) Introduction: an historical background sketch. In: Welford, A.T. (ed) *Reaction Times*. Academic Press, New York, pp.1–23.

Canfield, R.L., Smith, E.G., Brezsnyak, M.P. and Snow, K.L. (1997) Information processing through the first year of life: a longitudinal study using the visual expectation paradigm. *Monogr Soc Res Child Dev* **62**, 1–145.

Creamer, L.R. (1963) Event uncertainty, psychological refractory period and human data processing. *J Exp Psychol* **100**, 63–72.

Dane, S. and Erzurumluoglu, A. (2003) Sex and handedness differences in eye-hand visual reaction times in handball players. *Int J Neurosci* **113**, 923–929.

Dane, S. and Gumustekin, K. (2002) Correlation between hand preference and distance of focusing points of two eyes in the horizontal plane. *Int J Neurosci* **112**, 1041–1047.

Der, G. and Deary, I.J. (2006) Age and sex differences in reaction time in adulthood: results from the United Kingdom health and lifestyle survey. *Psychol Aging* **21**, 62–73.

Donders, F.C. (1868) On the speed of mental processes. In: Koster, W.G. (ed) Attention and performance II. *Acta Psychol* **30**, 412–431 (original work published in 1868).

Fitts, P.M. and Seeger, C.M. (1953) S–R compatibility: spatial characteristics of stimulus and response codes. *J Exp Psychol* **46**, 199–210.

Friston, K.J., Price, C.J., Fletcher, P. *et al.* (1996) The trouble with cognitive subtraction. *Neuroimage* **4**, 97–104.

Hick, W.E. (1952) On the rate of gain of information. *Q J Exp Psychol* **4**, 11–26.

Hultsch, D.F., MacDonald, S.W. and Dixon, R.A. (2002) Variability in reaction time performance of younger and older adults. *J Gerontol B Psychol Sci Soc Sci* **57**, 101.

Jackson, R.C. and Farrow, D. (2005) Implicit perceptual training: how, when, and why? *Hum Mov Sci* **24**, 308–325.

Kida, N., Oda, S. and Matsamura, M. (2005) Intensive baseball practice improves Go/No go reaction time, but not the simple reaction. *Brain Res Cogn Brain Res* **22**, 257–264.

Kornblum, S., Hasbroucq, T. and Osman, A. (1990) Dimensional overlap: cognitive basis for stimulus–response compatibility – a model and taxonomy. *Psychol Rev* **97**, 253–270.

Levy, J., Pashler, H. and Boer, B. (2006) Central interference in driving: is there any stopping the psychological refractory period. *Psychol Sci* **17**, 228–235.

Lieberman, H.R., Niro, P., Tharion, W.J. *et al.* (2006) Cognition during sustained operations: comparison of a laboratory simulation to field studies. *Aviat Space Environ Med* **77**, 929–935.

Lien, M.-C. and Proctor, R.W. (2002) Stimulus–response compatibility and psychological refractory period effects: implications for response selection. *Psychon Bull Rev* **9**, 212–238.

Linford, C.W., Hopkins, J.T., Schulthies, S.S. *et al.* (2006) Effects of neuromuscular training on the reaction time and electromechanical delay of the peroneus longus muscle. *Arch Phys Med Rehabil* **87**, 395–340.

Luchies, C.W., Schiffman, J., Richards, L.G. *et al.* (2002) Effects of age, step direction, and reaction condition on the ability to step quickly. *J Gerontol A Biol Sci Med Sci* **57**, M246–249.

McCrea, P.H. and Eng, J.J. (2005) Consequences of increased neuromotor noise for reaching movement in persons with stroke. *Exp Brain Res* **162**, 70–77.

Meyer, D.E. and Kieras, D.E. (1997a) A computational theory of executive cognitive processes and multiple-task performance: Part 1: basic mechanisms. *Psychol Rev* **104**, 3–65.

Meyer, D.E. and Kieras, D.E. (1997b) A computational theory of executive cognitive processes and multiple-task performance: Part 2: accounts of psychological refractory-period performance. *Psychol Rev* **104**, 749–791.

Miller, J.O. and Low, K. (2001) Motor processes in simple, go/no-go, and choice reaction time tasks: a psychophysiological analysis. *J Exp Psychol Hum Percept Perform* **10**, 266–289.

Mouelhi Guizani, S., Tenenbaum, G., Bouzaouach, I. *et al.* (2006) Information-processing under incremental levels of physical loads: comparing racquet to combat sports. *J Sports Med Phys Fitness* **36**, 335–343.

Newell, K.M. (1990) Motor skill acquisition. *Ann Rev Psychol* **42**, 213–237.

Pain, M.T. and Hibbs, A. (2007) Sprint starts and the minimum auditory reaction time. *J Sports Sci* **25**, 79–86.

Pashler, H. (1994) Dual-task interference in simple tasks: data and theory. *Psychol Bull* **116**, 220–244.

Posner, M.I. (1978) *Chronometric Explorations of Mind*. Erlbaum, Hillsdale, NJ.

Poulton, E.C. (1957) On prediction in skilled movements. *Psychol Bull* **54**, 467–478.

Rose, S.A., Feldman, J.F., Jankowski, J.J. and Caro, D.M. (2002) A longitudinal study of visual expectation and reaction time in the first year of life. *Child Dev* **73**, 47–61.

Savelsbergh, J.P., Van der Kamp, P., Williams, M.A. and Ward, P. (2005) Anticipation and visual search behavior in expert soccer goalkeepers. *Ergonomics* **48**, 1686–1697.

Silverman, I.W. (2006) Sex differences in simple visual reaction time. A historical meta analysis. *Sex Roles* **54,** 57–68.

Simon, J.R. and Small, A.M. (1969) Processing auditory information: interference from an irrelevant cue. *J Appl Psychol* **53**, 433–435.

Sperry, R.W. (1961) Cerebral organization and behavior. *Science* **133**, 1749–1757.

Stelmach, G. (1982) Information-processing framework for understanding human motor behavior. In: Kelso, J.A.S. (ed) *Human Motor Behavior: an introduction*. Erlbaum, Hillsdale, NJ, pp.63–91.

Telford, C.W. (1931) The refractory phase of voluntary and associative responses. *J Exp Psychol* **14**, 1–36.

Thomas, N.G., Harden, L.M. and Rogers, G.G. (2005) Visual evoked potentials, reaction times and eye dominance in cricketers. *J Sports Med Phys Fitness* **45**, 428–433.

Weinstein, S. and Graves, R.E. (2002) Are creativity and schizotypy products of a right hemisphere bias? *Brain Cogn* **49**, 138–151.

Section F

NEUROLOGICAL ISSUES

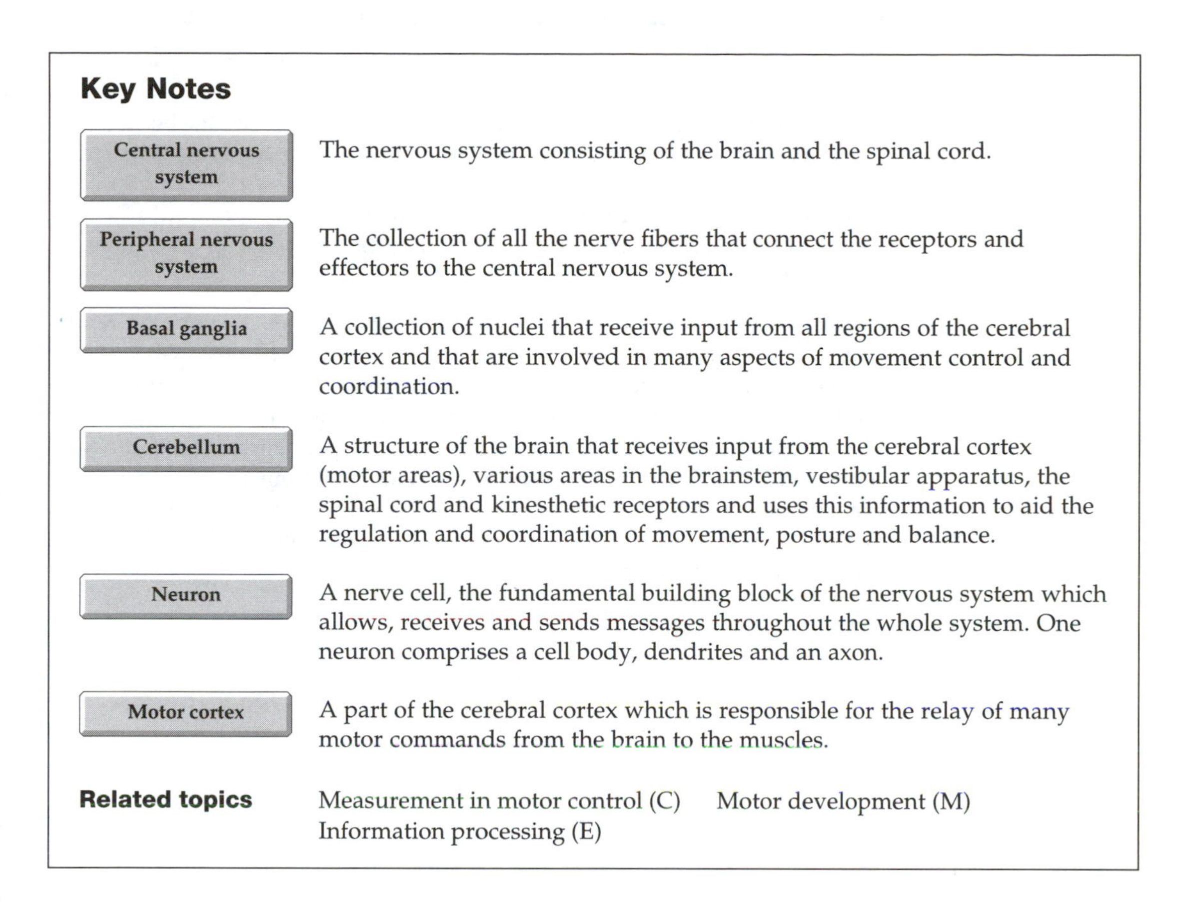

Key Notes

Central nervous system	The nervous system consisting of the brain and the spinal cord.
Peripheral nervous system	The collection of all the nerve fibers that connect the receptors and effectors to the central nervous system.
Basal ganglia	A collection of nuclei that receive input from all regions of the cerebral cortex and that are involved in many aspects of movement control and coordination.
Cerebellum	A structure of the brain that receives input from the cerebral cortex (motor areas), various areas in the brainstem, vestibular apparatus, the spinal cord and kinesthetic receptors and uses this information to aid the regulation and coordination of movement, posture and balance.
Neuron	A nerve cell, the fundamental building block of the nervous system which allows, receives and sends messages throughout the whole system. One neuron comprises a cell body, dendrites and an axon.
Motor cortex	A part of the cerebral cortex which is responsible for the relay of many motor commands from the brain to the muscles.

Related topics Measurement in motor control (C) Motor development (M)
Information processing (E)

Most humans are capable of producing movements that are unique, diverse and extremely complex. These movements can be deliberate such as returning a serve while playing tennis or playing the piano, or involuntary such as swallowing or correcting postural sway. All movements enable humans to interact with their environment and often have speed and/or accuracy requirements. In addition, movement requires complex coordination of many parts of the body and sometimes coordination between the body and other implements, such as a cricket bat. These movements have to be modifiable and adaptable, exhibiting flexibility to ensure that they can be performed in a myriad of ways in differing contexts. To support our need for such a versatile movement system capable of such an array of behavior, the human nervous system is designed as a highly organized, comprehensive, yet flexible system to complement the complex properties of movement itself. In the sections that follow, the nervous system and its main components will be examined and their role in the control of movement discussed.

The nervous system

The nervous system is the body's information gatherer, storage center and control system. Its overall function is to collect information, to analyze this information, and facilitate communication between different cells of the body to initiate appropriate responses to this information. The nervous system has mechanisms that allow the performer to respond to basic physiological states and external stimuli or to act purposefully in relation to objects. It also allows us to move quickly in response to danger, but also thoughtfully in more complex ways.

The nervous system has essentially four functions:

1. to gather information both from the outside world and inside the body
2. to transmit this information to the processing area of the brain and spinal cord
3. to process the information to determine the best response
4. to send information to muscles, glands and organs (effectors) so they can respond correctly.

To permit these sensory, integrative and motor functions the nervous system has the two specialist divisions: the central nervous system (CNS) and the peripheral nervous system (PNS) (*Fig. 1*). Together they help control every aspect of your daily life from breathing to performing a headstand. The basic unit of these systems is the neuron. Neurons enable us to interact with our environment and use electrical impulses to allow the communication of messages from one cell to another, moving only in one direction. The cell processes information from the sensory nerves and initiates an action within milliseconds, enabling the body to respond quickly.

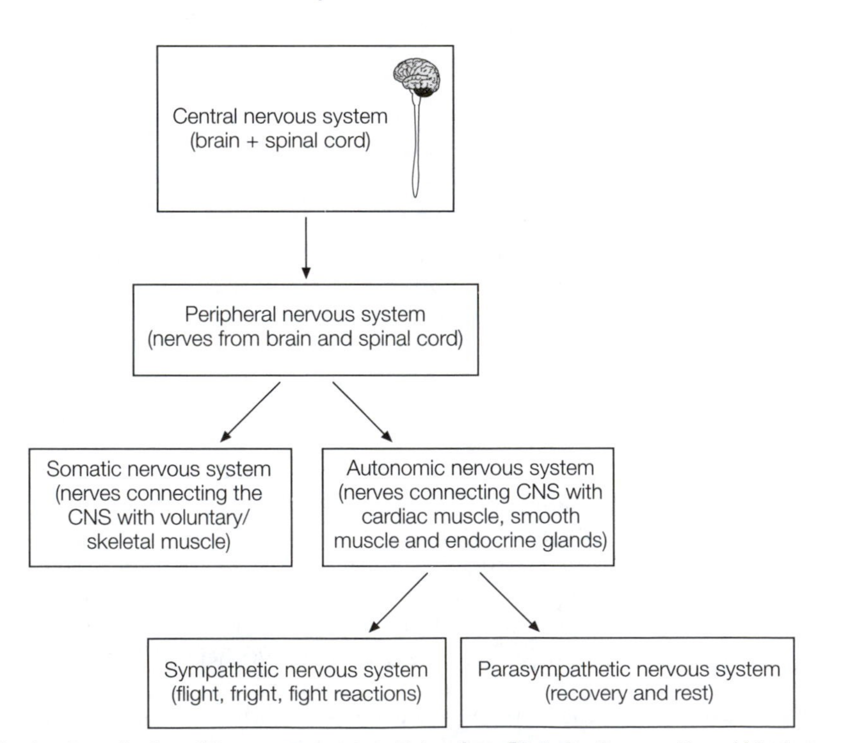

Fig. 1. Organization of the nervous system. Taken from Birch, K., George, K. and MacLaren, D. (2004) Instant Notes in Sport and Exercise Physiology. Taylor and Francis.

The neuron

The neuron is the functional unit of the nervous system and it is estimated that the brain contains 100 billion neurons. These specialized cells are essential for receiving and sending messages (information) throughout the entire neuromuscular system. Although human behavior is extremely complex this depends more on the fact that these cells form many precise anatomical circuits, rather than on the specialization of individual nerve cells. Neurons vary in size or shape; however all share the same basic architecture comprising three parts – a cell body (soma), dendrites, and an axon (*Fig.* 2). The cell body contains the nucleus, mitochondria and other organelles typical to a eukaryotic cell, and its function is to regulate the internal environment (homeostasis) of the neuron. The cell body usually gives rise to two processes: many dendrites and a single axon. The dendrites connect with other neurons and are the main apparatus for receiving incoming signals from other cells. In contrast, the axon is responsible for sending information to other neurons via electrical signals. These electrical signals are called action potentials and are rapid, transient, all-or-none nerve impulses that originate from dendrites close to the axon hillock. There are different types of neurons and their structure is dependent upon their function.

Action potentials are the means by which the brain receives, analyzes, and conveys information, which is determined by the pathway the signal travels to the brain. The transmission of the action potential is speeded up by a lipid membrane surrounding the axon called the myelin sheath (or principally Schwann cells in the PNS). The axon eventually divides into branches, at the end of which are presynaptic terminals; these allow communication with other neurons through the transmission of the electrical impulses. The passage of information from one neuron to another occurs via synapses, where the axon of one neuron comes into close proximity to, but does not touch or communicate anatomically with, another (postsynaptic) nerve cell. The space that separates two neurons is called the synaptic cleft. The electrical activity in the presynaptic neuron is transmitted across the synaptic cleft to the postsynaptic neuron through further electrical activity or a chemical mediator – a neurotransmitter.

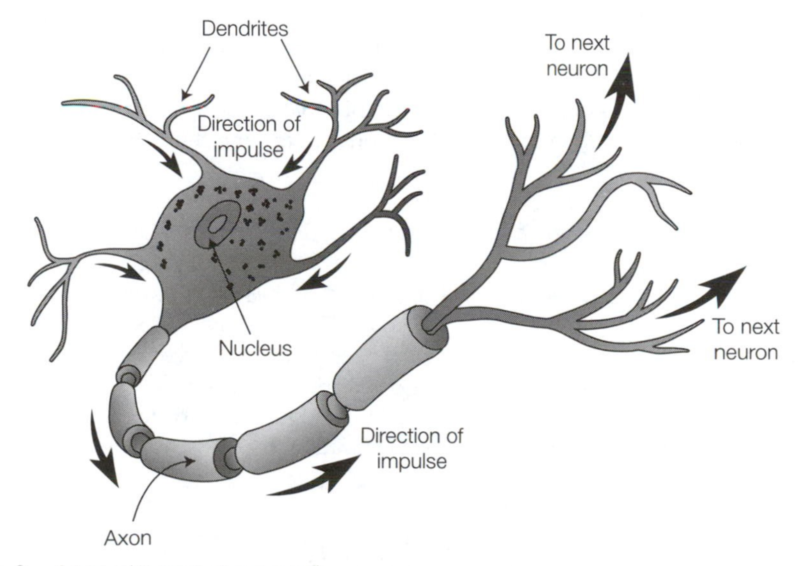

Fig. 2. A typical neuron or nerve cell.

When an action potential arrives at the terminal button of neuron, calcium (Ca^{2+}) channels open in the membrane and Ca^{2+} ions rush into the neuron. These ions cause the synaptic vesicles to fuse with the membrane, triggering the release of neurotransmitter molecules into, and across, the synaptic cleft. The neurotransmitter molecules, the best known being acetylcholine (ACh), then bind to the receptors of the postsynaptic membrane of the other neuron. These receptors detect the presence of the neurotransmitter and trigger either an excitatory or an inhibitory response in the postsynaptic membrane, dependent upon the nature of these specialized receptors. Thus the transmission of information between neurons is a culmination of both electrical and chemical activity.

The neurons of the spinal system are commonly classified into three major functional groups: sensory neurons, motor neurons and interneurons. Sensory neurons carry information picked up from the environment by the sense organs (receptors) to the appropriate parts of the nervous system for the purpose of both perception and movement coordination. Motor neurons carry information from the nervous system to muscles and glands (effectors), in response muscles contract and glands secrete. There are two main types of motor neurons; alpha and gamma motor neurons. Alpha motor neurons innervate extrafusal muscle fibers, which are responsible for carrying neural impulses to the appropriate muscle response synergies, allowing us to catch and throw a ball for example. Gamma motor neurons innervate intrafusal muscle fibers; these have a sensory role informing the brain of any differences between the intended muscle length and the actual muscle length, and are responsible for detecting and correcting errors when moving. Interneurons are found only in the CNS and consist of all nerve cells that are not specifically motor or sensory. These neurons are responsible for integrating the various sensory inputs that enter the CNS and permit the connection of multiple neurons with other multiple neurons. Generally, interneurons are divided into two classes: relay or projection interneurons and local interneurons. The former can convey signals between one brain part and another, while the latter process information within local circuits. Interneurons are particularly important for movement as they allow some muscles to be excited and contract while other are inhibited; this is called reciprocal inhibition. Reciprocal inhibition allows many muscular synergies to evolve and this permits many activities we take for granted in our daily lives.

Neurons do not act alone, but are joined together to form a complicated communication network that gives rise to the nervous system, which is divided into the central nervous system and the peripheral nervous system.

The central nervous system

The CNS is composed of the brain and spinal cord, which are surrounded by the skull and the vertebrae respectively. The role of the spinal cord is to carry messages from the body to the brain, where they are analyzed and interpreted. Response messages are then passed from the brain through the spinal cord and to the rest of the body. Thus the CNS is somewhat hierarchical in structure. The brain is responsible for both cognitive and motor control functions and the spinal cord is responsible for more routine or repetitive control functions. However, it is important to understand that in the performance of skilled movement both work cooperatively.

The spinal cord

The main function of the spinal cord is to act as an interface between the brain and the PNS. The spinal chord comprises both grey and white matter. The grey matter of the spinal cord consists mostly of cell bodies and dendrites. The surrounding white matter is made up of nerve fibers which are bundled together to form tracts that carry information up and down the spinal cord. Thirty-one pairs of spinal nerves, part of the PNS, emerge from the spinal cord. Each nerve has a dorsal root, which carries afferent sensory information to the brain, and a ventral root, which carries efferent or motor information from the brain. The ventral root mostly comprises the axons of alpha motor neurons, whose cell bodies are located in the anterior part of the spinal cord. The dorsal root comprises both the axons of the sensory neurons and their cell bodies, which are clustered together to form the dorsal root ganglion. In addition to carrying electrical impulses to and from the brain, the spinal cord is also involved in reflexes that do not immediately involve the brain, these are often referred to as 'non-conscious' spinal reflexes (Latash, 1998).

Reflexes

A reflex is an involuntary coordinated pattern of contraction and inhibition of muscle groups and is elicited by peripheral stimuli; examples of such reflexes are blinking or sneezing. Reflexes produce a rapid motor response to a stimulus, the motor response being either contraction or inhibition of the muscle fibers. Reflexes are classified as either monosynaptic or polysynaptic, involving either one, or several neurons respectively, and the pathway a single reflex travels is called the reflex arc.

The best example of a monosynaptic reflex within the spinal cord is the myotatic reflex or the muscle-stretch reflex. The stimulus for the stretch reflex is excessive stretch on the muscle detected by the muscle spindle. For example, if the weight on the arm is suddenly increased, the arm is pulled down and the sensory neurons from the intrafusal fibers detect the change in length. This results in activity in these neurons and a nerve impulse being sent to the dorsal root of the spinal cord and then directly to an alpha motor neuron via a single synapse. The alpha motor neurons then transmit the nerve impulse back to the extrafusal muscle fibers of the muscles of the arm causing the muscle to contract and limb position to be restored.

In a polysynaptic reflex, sensory and motor neurons do not synapse directly, but the impulses there are involved within the reflex arc, the longer the reflex loop time, or the time taken from the presentation of the external stimuli to the time a response is recorded in the muscle fibers. The simplest example of a polysynaptic reflex or a flexion reflex is the withdrawal of a limb from painful stimulus via a process called reciprocal inhibition. The pain receptor causes the sensory neuron to fire; this then proceeds via the dorsal root to synapse with the interneuron(s) in the spinal cord. The excitatory interneuron then synapses with a motor neuron causing the motor neuron in the ventral root to fire. This excitation causes subsequent contraction of the flexor muscle. At the same time, and just as important, the extensor muscle is stretched and inhibitor interneurons prevent the extensor muscle from firing, to allow the limb to move away from the stimulus.

More complex reflexes can coordinate the activity of several muscle groups, sometimes on different sides of the body. The crossed extensor reflex often functions in conjunction with the flexion reflex to help maintain postural stability or help a person push away from a painful stimulus. For example, interneuronal connections involved in the crossed extensor reflex cause extensor muscles in the opposite limb to contract at the same time as the flexor muscles remove the limb from a painful stimulation. This provides an excellent example of how nerve impulses pass across the spinal cord from one side of the body to another and not just to and from the spinal cord at a given segment level or up and down the spinal cord.

Research Highlight: Dragansk *et al.* (2004)
Dragansk *et al*. (2004) conducted a fascinating study which aimed to examine if the adult human brain alters in response to environmental demands. This study involved a cascade juggling task which 24 subjects took part in, divided into two groups: jugglers and non-jugglers (neither group could juggle before the experiment). Brain scans were performed pre, post, and 3 months after the intervention. Comparisons between the groups showed no significant differences in grey matter between the juggling and non-juggling group pre-intervention. However post-intervention there was a significant difference, with the juggler group having a significant transient bilateral expansion in grey matter in the mid-temporal area and in the left parietal sulcus compared to the non-juggler group. This expansion was not as significant after 3 months, and the non-juggling group showed no changes. These results challenge the traditional view that the anatomical structure of the human brain does not alter after skill acqusition and that cortical plasticity is associated with functional rather than anatomical changes.

Control of movement by the brain

Nerves reach from your brain to your face, eyes, ears, nose, mouth and spinal cord and from the spinal cord to the rest of your body. Afferent (sensory) nerves gather information from the environment and send that information to the spinal cord, which in turn speeds a message to the brain. The brain then interprets that message and fires off a response which the efferent nerves (motor) deliver to the rest of the body.

The brain is a highly specialized structure at the end of the spinal column, that takes the form of a distorted hollow tube filled with cerebrospinal fluid. The brain is highly complex serving many functions. In this section we will examine those areas of the brain that have been identified as having a significant role in the control of movement (*Fig. 3*).

The cerebral cortex

The cerebral cortex is the executive suite of the nervous system and it enables us to communicate, perceive, and produce voluntary movement. Three types of functional areas can be found in the cerebral cortex: motor areas, sensory motor areas, and associated areas. The cerebral cortex is the largest part of the human brain, associated with higher brain functions such as thought and action. The cerebral cortex is divided into two hemispheres which, although they look symmetrical, have somewhat different functions. The left and right hemispheres of the cerebral cortex are connected by the corpus callosum, a large bundle of interconnecting nerve fibers. The cerebral cortex can be divided into four 'lobes': the frontal lobe, parietal lobe, occipital lobe and temporal lobe, with each lobe being represented in each hemisphere.

The frontal lobe houses the structures of major importance for the control of voluntary movement. These include the motor cortex, which projects directly to the muscles, and the premotor and supplementary motor cortex. These areas

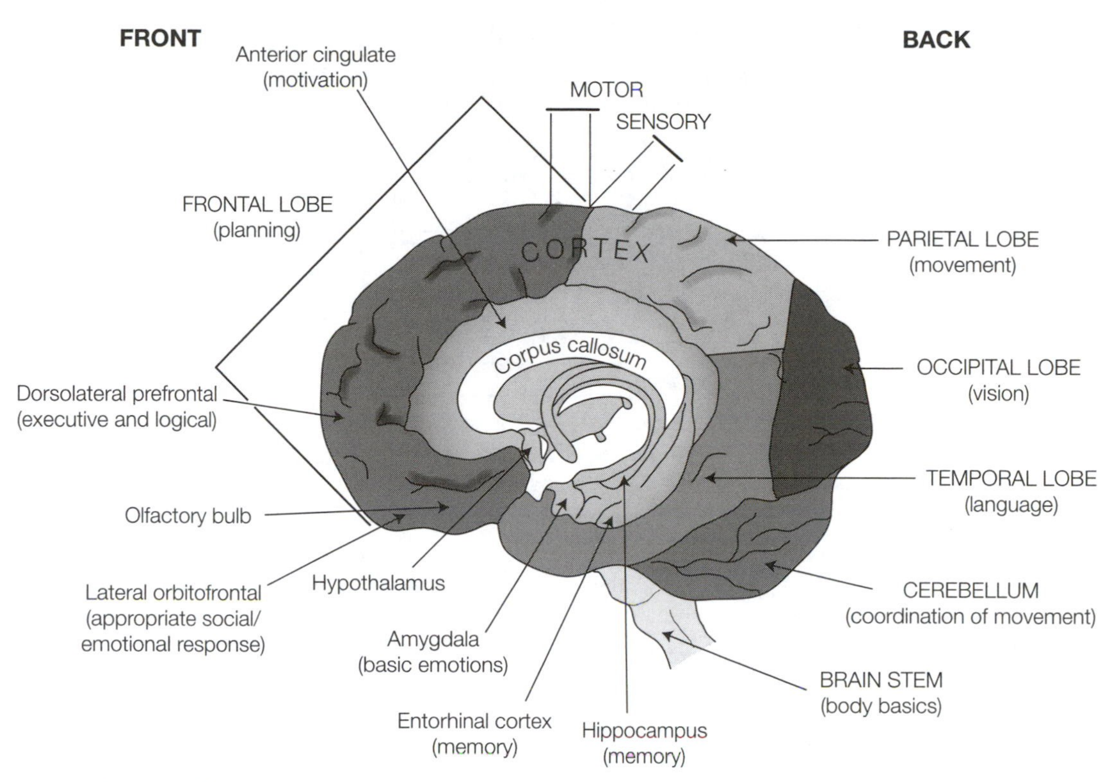

Fig. 3. Structures of the brain.

interact with the parietal cortex, which receives kinesthetic information essential for appropriate motor control. The motor areas also receive input from thalamocortical projections from the thalamus. The thalamus, considered to play a key role in sensorimotor integration, receives information from the other main CNS regions essential for motor control, namely the spinal cord, basal ganglia and cerebellum, which identify where we want to move, plan the movement and execute our actions (Ghez *et al.*, 1991). The relationship between the motor cortex and the other CNS regions associated with motor control is shown in *Fig. 4*.

The motor cortex can be regarded as a keyboard where each key represents a different area of the body. The keyboard is played by adjacent parts of the frontal lobe, particularly the supplementary motor area and the premotor cortex. Each of these structures is intimately involved in the production and control of skilled movement. The planning of complex behavior probably occurs in the prefrontal cortex; then the premotor cortex, along with the supplementary motor area receive sensory information from the parietal and temporal lobes. So, at the simplest level, the parietal and temporal lobes analyze sensory information, the prefrontal cortex and supplementary area decide what action (behavior) is to be taken, and then the motor cortex executes the instructions. Damage to the cerebral cortex can result in apraxia. Apraxia is characterized by inability to perform a skilled or learned act that cannot be explained by an elementary motor or sensory deficit or language-comprehension disorder despite having the desire and the physical ability to perform the act (Zadikoff and Lang, 2005). Other research has shown that the damage to the left or right hemispheres results in differing motor impairment. Steenbergen *et al.* (2004)

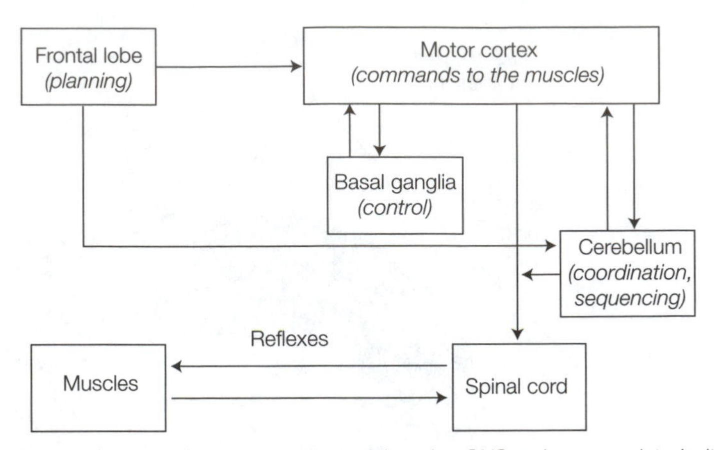

Fig. 4. Linkages between the motor cortex and the other CNS regions associated with motor control.

found that subjects with left brain damage have difficulties with planning a task. Similarly Haaland (2006) demonstrated that left frontoparietal circuits control limb praxis and motor sequencing.

Each part of the motor cortex and its associated area controls specific muscles and muscle groupings. Thus, all muscles are topographically represented in the brain. The primary motor cortex (Brodman's area 4) contains a complex map of the body or motor homunculi (*Fig. 3*) in which each region of the body is represented, and there is predominantly contralateral control; the right motor cortex coordinates muscles in the left side of the body, and vice versa. The relationship between cortical 'space' in the motor strip and body region is not proportionate: there is over-representation for the regions involved in fine motor control, such as the hands and fingers and the mouth area of the face (Penfield and Rassmussen, 1952). At the same time there is under-representation of less 'movement-critical' regions such as the back and the top of the head, the trunk and the upper limbs.

The motor cortex has two principal means of relaying commands to the many muscles of the body; these are through the pyramidal tract and the extrapyramidal tract (the main attributes and functions of which are shown in *Box 1*).

Box 1. Major motor pathways and their functions

Pyramidal

- Precise, discrete movements, e.g. single finger
- Unilateral
- Cortical input; mainly from primary motor cortex
- Damage leads to paralysis.

Extrapyramidal

- Less precise, general
- Bilateral
- Cortical input; wide areas of cortex
- Subcortical input; basal ganglia, cerebellum
- Damage leads to abnormal movements.

The pyramidal tract is the more direct of the two routes and carries messages for voluntary motor movement to motor neurons in the brainstem and spinal cord. This tract is direct and monosynaptic, meaning that the axons of its neurons do not synapse with other cells until they reach their final destination in the brainstem or spinal cord. These direct connections between the cortex and the lower motor neurons allow messages to be transmitted very rapidly from the CNS to the periphery.

The pyramidal system is highly vulnerable to damage, and the typical result of this damage is partial (hemiparesis) or complete paralysis (hemiplegia). As the pyramidal tract is essentially an excitatory system, it exists to tension muscles, and when it is damaged that tension simply disappears and the muscle eventually wastes away.

The extrapyramidal motor system is a highly complex system of major ganglia and tracts responsible for the control of involuntary muscle excitation. This tract allows nerve impulses from the motor cortex to reach spinal level through a range of pathways via the basal ganglia, thalamus and cerebellum. Outputs are primarily inhibitory and damage to or a lesion of the extrapyramidal tract results in spastic paralysis and increased reflexes.

The basal ganglia

The basal ganglia are located deep within layers of the cerebral hemispheres near the thalamus. They consist of five interconnected nuclei and are considered to be part of the extrapyramidal system. The major output nuclei are the internal globus pallidus and the substantia nigra, which both project to the ventral thalamus. In addition, two of the nuclei, the caudate nucleus and the putamen form the striatum, which is primarily involved in input from the cerebral cortex and thalamus.

The basal ganglia are involved in monitoring and controlling the activity of the motor cortex, however to date the precise function remains elusive. Some suggest they are involved in the control of slow movements and those that require the retrieval and initiation of movement plans such as handwriting, and some suggest that they are involved in the learning and control of tasks that become automatic such as driving a car.

Problems with the basal ganglia can have a dramatic effect on movement. Degeneration of the release of the natural neurotransmitter dopamine from the substantia nigra causes the degenerative Parkinson's disease. The resulting deficiency of dopamine results in excitatory input from the basal ganglia to the motor cortex, leading to hypokinesia or braykinesia (slowing of movements) and akinesia (difficulty in initiating movement). Postural movements are also impaired, probably because of the reduced excitatory input to the ventromedial system. In general, Parkinson's patients typically demonstrate a range of motor symptoms such as shuffling, uncertain gait, limb tremor and high muscle stiffness.

In stark contrast, the hereditary disease Huntingdon's is associated with hyperkinesia or uncontrolled, jerky, excessive, involuntary movements particularly evident in the limbs and/or facial muscles. These symptoms are thought to be due to a loss of projections between the striatum and globus pallidus.

The cerebellum

The second largest brain structure, the cerebellum, is crucial for motor control, and lesions of this area produce devastating effects on our ability to perform movement, from the very simple such as reaching for a glass of water to the very elegant such as a pirouette. The cerebellum, or 'little brain', is similar to the cerebrum in that it has two hemispheres and a highly folded surface or cortex. Like the basal ganglia, it is considered to be part of the extrapyramidal system. The cerebellum comprises three lobes: the anterior, posterior and flocculondular lobes, and three pairs of deep nuclei: the fastigial, the interposed and the denate nuclei (Latash, 1998).

The cerebellum receives input from the cerebral cortex (motor areas), from various areas in the brainstem, the vestibular apparatus, the spinal cord and kinesthetic receptors. The afferent fibers of the cerebellum (which are more prevalent than the efferent fibers; Latash, 1998) are of two types. The climbing fibers originate in the medulla (located in the brainstem), and the mossy fibers originate primarily from the brainstem and spinal cord. The major outputs of the cerebellum via its output fiber (Purkinke cell) are to the thalamus and the brainstem. These inhibitory cells are the largest neurons in the brain (Latash, 1998).

The principle functions of the cerebellum appear to be the regulation of muscle tone, the coordinated smoothing of movement timing, and learning permitting the regulation and coordination of movement, posture and balance. The cerebellum is thought to act as a comparator, adjusting its motor responses by comparing the intended output with sensory signals and updating the movement commands if they deviate from the intended path. Morton and Bastian (2006) conducted an experiment looking at locomotor adaptability in an unpredictable environment. They were interested in examining which structures control different types of adaptation and which specific mechanisms make appropriate adjustments. They used a splitbelt treadmill to test cerebellar contributions to two different forms of locomotor adaptation in humans. It was found that the cerebellum seems to play an essential role in predictive but not reactive locomotor adjustments (see Further reading, below).

The cerebellum also modulates the force and range of movements, and the importance of this structure for this becomes strikingly obvious when observing patients who have suffered trauma to the cerebellum. Despite their understanding of the movement goal, patients with cerebellar lesions are unable to apply the movement dynamics necessary to implement planned actions (Rose and Christina, 2005). Individuals with cerebellar dysfunction appear unsteady, and complex sequential actions are extremely difficult to complete. Hypermetria is often common is patients with cerebellar lesions, for example when reaching for an object the patient may overshoot the desired object or widen their hand more than is necessary when approaching an object they wish to pick up. Given the fact that the cerebellum is primarily responsible for coordination and regulation of action, it is not surprising that motor control is affected in this way.

The peripheral nervous system

The PNS serves as a bridge between the environment and the CNS and consists of the neurons that connect the brain and the spinal cord with the muscles and the glands of the body, while the sensory system includes all the neural tissue outside of the CNS. The PNS is responsible for providing sensory or afferent information to the CNS and carrying motor, or efferent commands out to the body's tissues. There are two major subdivisions of the PNS motor pathways: the somatic and autonomic motor systems.

The somatic nervous system is the part of the PNS associated with voluntary control of body movements through the action of skeletal muscles, and also reception of external stimuli. The somatic nervous system consists of afferent fibers that receive information from external sources and enter the spinal cord in the dorsal root ganglion, and efferent fibers, which are derived from the ventral root of the spinal cord and are responsible for muscular contraction.

The autonomic nervous system consists of sensory neurons and motor neurons that run between the CNS (especially the hypothalamus and medulla oblongata) and various internal organs such as the heart, lungs and glands. It is responsible for monitoring conditions in the internal environment and bringing about appropriate changes in them. The contraction of both smooth muscle and cardiac muscle is controlled by motor neurons of the autonomic system.

The actions of the autonomic nervous system are largely involuntary (in contrast to those of the somatic system). It also differs from the somatic system is using two groups of motor neurons to stimulate the effectors instead of one. The first, the preganglionic neurons, arise in the CNS and run to a ganglion in the body. Here they synapse with postganglionic neurons, which run to the effector organ (cardiac muscle, smooth muscle or a gland).

The efferent or motor autonomic fibers are described as either sympathetic or parasympathetic. These two systems often innervate the same organs and have roughly opposite roles. The sympathetic nervous system is associated with activity, expending energy, and excitement. Activity in the sympathetic division causes an increase in heart rate, release of adrenaline and a rise in blood sugar. The parasympathetic system is responsible for the storing of energy, digestion etc. Activity in the parasympathetic division slows heart rate via the vagus nerve, the largest cranial nerve in the human body.

Conclusion

The CNS (brain and spinal cord) has ascending sensory pathways and descending motor pathways in the control tower of the human system. It both receives and carries information, which is vital to our ability to move and survive. It is responsible for overseeing and monitoring the activation of all sectors of the human body. Over time, the depth of our understanding of the CNS and its workings has increased. Changes in technology have meant that we are still discovering how the CNS, especially the brain, function. An understanding of the CNS is crucial for understanding motor control, learning and development. A number of studies (see below) that have been conducted recently have further added to our knowledge of the CNS; this will continue as researchers are better able, via techniques such as functional magnetic resonance imaging, to explore more complex motor skills.

Further reading

Kandel, E.R., Schwartz, J.H. and Jessel, T.M. (2000) *Principles of Neural Science*, 4th Edn. McGraw-Hill, New York, pp.48–89.

Latash, M.L. (1998) *Neurophysiological Basis of Movement*. Human Kinetics, Champaign, IL.

Mikheev, M., Mohr, C., Afanasiev, S., Landis, T. and Thut, G. (2002) Motor control and cerebral hemispheric specialization in highly qualified judo wrestlers. *Neuropsychologia* **40**,1209–1219.

Morton, S.M. and Bastian, A.J. (2006) Cerebellar contributions to locomotor adaptations during splitbelt treadmill walking. *J Neurosci* **26**, 9107–9116.

Shumway-Cook, A. and Woollacott, M. H. (2001) *Motor Control: theory and practical applications*, 2nd Edn. Lippincott, Williams and Wilkins, Philadelphia, pp.163–191.

References

Dragansk, B., Gase, C., Busch, V. *et al.* (2004) Neuroplasticity: changes in grey matter induced by training. *Nature* **427**, 311–312.

Ghez, C., Hening, W. and Gordon, J. (1991) Organization of voluntary movement. *Neurobiology* **1**, 664–671.

Haaland, K.Y. (2006) Left hemisphere dominance for movement. *Clin Neuropsychol* **20**, 609–622.

Latash, M.L. (1998). *Neurophysiological Basis of Movement*. Human Kinetics, Champaign, IL.

Morton, S.M. and Bastian, A.J. (2006) Cerebellar contributions to locomotor adaptations during splitbelt treadmill walking. *J Neurosci* **26**, 9107–9116.

Penfield, W. and Rasmussen, T. (1952) *The Cerebral Cortex of Man.* Macmillan, New York.

Rose, D.J. and Christina, R.W. (2005) *A Multilevel Approach to the Study of Motor Control and Learning*, 2nd Edn. Pearson, San Francisco.

Steenbergen, B., Meulenbroek, R.G.J. and Rosenbaum, D.A. (2004) Constraints on grip selection in hemiparetic cerebral palsy: effects of lesional side, end-point accuracy, and context. *Brain Res Cogn Brain Res* **19**, 145–159.

Zadikoff, C. and Lang, A.E. (2005) Apraxia in movement disorders. *Brain* **128**, 1480–1497.

Sensory contributions to control

Key Notes

Exteroceptive — Information from the external environment, for example vision or audition.

Proprioception — Sensation and perception of head, trunk and limb movement.

Interoceptors — Receptors that provide information about the state of functioning of internal organs, but which are not typically associated with or primarily responsible for movement control (e.g. baroreceptors which tell us about blood pressure).

Deafferentation — The elimination or interruption of sensory nerve impulses via surgery or injury.

Related topics

Theories of control (D)
Neurological issues (F)
Implications for practice (K)

When you perform any movement, from everyday skills such as opening a door to more complex actions such as crossing a road, gymnastic skills as seen in *Fig. 1*, or catching a ball, you use information detected by sensory receptors located in various parts of the human body. For all movements there are two types of environmental information we use. Information from the external environment is called exteroceptive information, the most important of which is vision. Information from the internal environment (the human body) is called kinesthetic information or proprioception (body awareness). Both visual information and proprioception are key features of any theory of motor control (see Section D). For example, in open-loop movement control, sensory information is used in movement planning to determine what motor commands the brain should send to the muscles. Sensory information is also used during movement execution, often to guide movements as they take place. This section considers what and how sensory information is used when deciding to move and in controlling movement.

Proprioception and movement

Proprioception is the sensory faculty of being aware of the *position* of our head, trunk and limbs. We can use this kind of information to 'fine tune' our movements. Proprioceptive information is conveyed in the same manner as kinesthetic information; the difference between the two of course is that kinesthesis deals with *motion* of the limbs, rather than their positioning. Kinesthesis allows performers to know whether a movement has been performed correctly by sensing how the movement 'felt' to perform it, in addition to being able to observe its effect.

Fig. 1. In order to know our position, orientation and speed when moving we use a range of exteroceptive information and proprioception.

Proprioceptors are located on the nerves, muscles, tendons, joints and inner ear and provide intrinsic information about the movement and balance of the body during the performance. They provide information to the brain regarding the adjustment of posture and movement, and influence the responses required for the body to correct imbalance due to tension in muscles or ligaments caused by a twisted foot for example. The spinal cord and brain centers are made aware of the situation and respond accordingly. Below are brief descriptions of four different proprioceptors: the muscle spindle, Golgi tendon organs, the vestibular apparatus and joint receptors. In summary the sensors listed in *Box 1* provide proprioceptive information about the movement of the body.

Box 1. Receptors that provide proprioceptive information

- *Tendon receptors*: Golgi tendon organ – muscle tension
- *Muscle receptors*: muscle spindle – stretch receptors
- *Joint receptors*: varied
- *Cutaneous receptors*: skin receptors.

By using these sensors we can tell exactly what position our limbs are in. Whether walking or swimming we know what angle the joints have reached and how much force the muscles are applying (*Fig. 2*).

Fig. 2. The swimmer needs to know the position of her limbs and the angle of the joints.

Muscle spindles

Muscle spindles are found in most skeletal muscles but are particularly concentrated in muscles that exert fine motor control (e.g. the small muscles of the hand) and large muscles which are rich in slow twitch muscles fibers. As its name implies, a muscle spindle is a spindle-shaped organ composed of a bundle of modified muscle fibers innervated by both sensory and motor axons. Muscle spindles lie in parallel in between regular muscle fibers, the distal ends being attached to the connective tissue within the muscle. Muscle spindles are stretch receptors, and when the muscle is stretched the spindle increases the discharge rate of the afferent fibers. This sends a message to a motor neuron in the spinal cord, which in turn relays a message to the muscle causing a contraction and thus shortens the muscle (e.g. the knee-jerk reflex, see Section M).

The Golgi tendon organ

The Golgi tendon organ is another type of proprioceptor or stretch receptor found in skeletal muscle, which provides information about changes in muscle tension. Golgi tendon organs are located at the junction of tendon and skeletal muscle and are formed from the terminals of Ib afferent fibers. The sensory endings of Golgi tendon organs are arranged in series with the muscle, in contrast to the parallel arrangement of the muscle spindles. Because of their arrangement, Golgi tendon organs can be activated by either muscle stretch or muscle contraction, with the latter being a more effective stimulus. The actual stimulus that activates the Golgi tendon organ is the force that develops in the tendon containing the Golgi tendon organ. Thus, the difference between a muscle spindle and a Golgi tendon organ is that the latter signals force, while the former signals muscle length and the rate of change in muscle length. The job of the Golgi tendon organ is to act as a brake against excessive contractions by inhibiting the motor neurons in the spinal cord.

Vestibular apparatus

The vestibular apparatus is found on each side of the head in the temporal lobe of the inner ear. It is composed of three semicircular ducts and two otolith organs. The semicircular ducts include the horizontal, superior and posterior ducts, the otolith organs include the utricle and the saccule; in addition an ampulla is found on each semicircular duct. The semicircular ducts all connect with the utricle and the utricle is joined to the saccule (*Fig. 3*).

The vestibular system detects angular and linear accelerations of the head. Signals from the vestibular system trigger eye and head movements to provide the retina with a stable visual image. In addition, signals from the vestibular apparatus allow the body to make adjustments in posture to maintain balance.

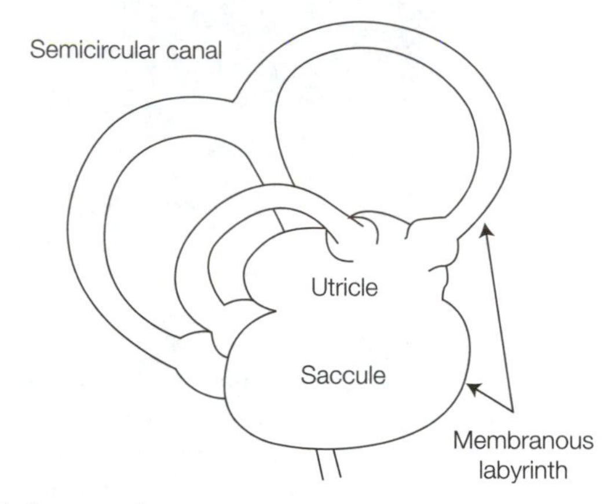

Fig. 3. The vestibular apparatus.

Joint receptors

Joints are associated with several different kinds of sensory receptors, including mechanoreceptors, nociceptors and ergoreceptors. Nociceptors in muscles respond to pressure applied to muscle and respond to hyperextension and hyperflexion, although many nociceptors fail to respond to joint receptors in normal conditions. Joint nociceptors are innervated by finely myelinated or unmyelinated primary afferent fibers. Joint mechanoreceptors include rapidly adapting mechanoreceptors – pacinian corpuscles, which respond to transient mechanical stimuli, including vibration – and slowly adapting receptors – Ruffini endings, which respond best to extreme movements of a joint and signal pressure or torque applied to the joint. Joint mechanoreceptors receptors are innervated by medium-sized afferent fibers.

The role of proprioception

Most theories of movement execution can be classified as open-loop or closed-loop theories (see Section D). In open-loop movement control, sensory information is used in movement planning to determine what motor commands the brain should send to the muscles, but once these commands are received, movement takes place without the controller receiving sensory feedback from the moving limb. The alternative is to use sensory information to monitor the progress of movements as they are executed, which is essentially a closed-loop system. Closed-loop systems make use of proprioceptive feedback to produce coordinated, controlled movements. Over the years experimenters have used a variety of ways to help us understand the role of this information for successful movement. Proponents of the dynamical systems approach to control would also consider how proprioceptive information would be used to control multiple degrees of freedom of control. Proprioceptive information would therefore be used to form and adjust coordinative structures.

Using surgical deafferentation to understand proprioception

The role of proprioception information for control of movement has been emphasized using a deafferentation model. Deafferentation refers to the elimination or interruption of sensory nerve impulses. Surgical deafferentation involves cutting or removing afferent neural pathways, which effectively stops information about the limb movement direction being sent to the central nervous system (CNS). Research has shown that when a single limb is deafferented in motorically mature monkeys there is loss of effective purposive use of that extremity in a free situation (Taub and Berman, 1963; Taub *et al.*, 1977). The work showed that the limb was not paralyzed, but that the limb moved frequently albeit rather randomly. In addition, researchers reported that although the monkeys were found to have preserved gross motor control, they showed severe difficulties in using their limb proficiently (Bossom, 1974; Knapp *et al.*, 1963; Taub and Berman, 1963). Just a little over 10 years later, Bizzi and coworkers showed that monkeys who were surgically deafferented could make accurate pointing movements even when proprioceptive feedback was no longer available, although it is important to note that the monkeys were trained prior to deafferentation to complete the same task (Bizzi and Polit, 1979; Polit and Bizzi, 1978).

For ethical reasons, deafferenting humans for experimental work alone is not possible. In an attempt to corroborate the results found in monkeys, researchers examine people who have had joint replacement surgery. Joint replacement surgery results in limited proprioception due to the absence of joint receptors. Fuchs *et al.* (1999) examined the differences in angle reproduction capability after total knee arthroplasty (TKA) and compared this to the patient's contralateral leg and a control group. The participants were asked to reproduce a 30 or a 60 degree angle from a starting position of 0 or 90 degrees either with or without vision. Fuchs *et al.* (1999) found some reduced proprioceptive capabilities in both the operated and contralateral leg of the TKA patients and some significant differences between the TKA and control group. For example, when vision was not available the TKA patients showed a larger deviation from the required angle of 30 degrees whether they started at 0 or 90 degrees.

Non-surgical deafferentation and proprioception

Deafferentation can also arise without surgery. For example, sensory polyneuropathy results in a loss of all sensory information (except for pain and temperature) due to the loss of or demyelination of large sensory fibers. The efferent fibers, however, are usually left intact. Mazzaro *et al.* (2005) examined the effect of dorsiflexion enhancements and reductions applied to patients with demyelination of large sensory fibers and age-matched controls during the stance phase of the gait cycle. This experimental procedure was designed to examine the contribution of these large-diameter sensory fibers to the adaptation of soleus muscle activity after small ankle trajectory modifications during human walking. In healthy subjects, the soleus electromyogram (EMG) gradually increased or decreased when the ankle dorsiflexion was, respectively, enhanced or reduced. Similarly, the soleus EMG activity in the patients increased during the dorsiflexion enhancements; however, the velocity sensitivity was decreased compared with the healthy volunteers. In addition, when dorsiflexion was reduced, the soleus EMG activity was unchanged. These data showed how important sensory feedback in the form of proprioception is for the continuous adaptation of soleus activity during the stance phase of walking.

Another well-established method for examining the role of proprioception is the tendon vibration technique, which effectively distorts proprioceptive feedback. This stimulation is known to induce a sensation of movement in a motionless limb, corresponding to the stretching of the stimulated muscle. In addition, when applied online, this technique modifies the sensorimotor function of the proprioceptive input (Hay *et al.*, 2005). This type of stimulation has been used to show the involvement of proprioception in regulation of vertical posture (Hay *et al.*, 1996), or whole-body movements (Quoniam *et al.*, 1995) and to evaluate the role of proprioception in speech and oral motor control (Loucks and De Nil, 2001).

Research Highlight: Does strapping enhance proprioception? (Simoneau *et al.*, 1997)
Simoneau *et al.* (1997) examined the assumption that taping joints to prevent injuries does so due to the increased proprioception that it provides through stimulation of cutaneous mechanoreceptors. Two strips of athletic tape (12.7 cm) were applied over the skin of the ankle in a single group of 20 healthy males. Using a randomized repeated measures design, ankle joint movement and position perception for plantar flexion and dorsiflexion were measured using a specially designed apparatus. Analyses of data indicated that under the non-weightbearing conditions, taping significantly improved the ability of the subjects to perceive ankle joint position but not movement perception. In the weightbearing condition, the use of tape did not significantly alter the ability of the subjects to perceive ankle position or movement perception. The authors concluded that increased cutaneous sensory feedback provided by strips of athletic tape applied across the ankle joint of healthy individuals can help improve ankle joint position and perception, but only in non-weightbearing situations.

Exteroceptive information

Exteroreceptive information consists of all the sensory information that we make use of arising from outside of the human body. For example, when crossing the road we make use of both visual and auditory information to decide when it is safe to do so. Although we make use of many rich sources of sensory information to control and alter our movement patterns in response to perturbations, it is the visual system that is uniquely positioned to provide information on the static and dynamic features of our environment (Patla, 1997).

Vision and motor control

Many of the motor skills humans perform require the use of vision to accurately produce the movement. Although many motor skills can use other sources of information, such as proprioception or tactile senses, vision is the dominant sensory system used by humans in the performance of motor skills, even when it is not the best source of information (Reeve *et al.*, 1986).

The visual field is believed to extend 160 degrees vertically and 200 degrees horizontally, and information from different segments of the visual field is picked up by the two components of the visual system. One component, central vision (sometimes referred to as focal or foveal), is limited and can only process information in the foveal region (approximately 2–5 degrees). Central vision allows one to see clearly in order to recognize objects and read displays; however, since it requires conscious thought, it is a relatively slow process (200 ms) so it cannot be used in some movements (e.g. response to a baseball pitch) (Keele and Posner, 1968).

The detection of information outside of the fovea region occurs by means of peripheral vision (also referred to as ambient vision). Peripheral vision is primarily responsible for spatial orientation; it is a subconscious function independent of focal vision whose primary role is to orient an individual in the environment. For example, you can occupy focal vision by reading this book (a conscious action), while simultaneously obtaining sufficient orientation cues with peripheral vision to walk to your next lecture (a subconscious function). Peripheral vision is important in many sports. For example, a defender may have their eyes fixed on the player with the ball; however, at the same time they will perceive a player running down the field to perhaps receive the pass, but will not likely be able to determine detail (opponent's identity or the quality of coverage by a team-mate); *Fig. 4*.

Fig. 4. When performing, a mass of visual information about the position of opponents and team-mates is available.

Optic flow

Another important issue related to how visual information contributes to the control of movement is concerned with using visual information in a dynamic or moving environment. Gibson (1966, 1979) suggested a living system and its environment are constantly engaged in energy transactions, and an athlete, for example, will rely on information in the environment to support the coordination of actions. Gibson suggested that visual information can be directly extracted from the environment and that, as a direct relationship is believed to exist between perception and action, there is no need for any intervening cognitive processes to render it meaningful.

Gibson (1966, 1979) argued that the layout of the environment is specified by the optical array that it emits, that is the light that is perceived by an individual. As we move in the environment the pattern of light on the retina varies depending on the layout of the environment, the level of illumination and the way in which we are moving. For example, if a person moves forward in a stable environment then the optical array will undergo an outward flow pattern. The optic array or optic flow contains information about the direction and the magnitude and velocity of a movement (Lappe *et al.*, 1999). Perhaps the best developed of the optic invariants that the human performer becomes attuned to with experience is 'tau' (denoted as the inverse rate of dilation of the image of an approaching object on the perceiver's retina; Lee, 1976). Tau is proposed as an optic variable that may be used to specify the time-to-contact of impending collision of human with object or surface. However, Tresilian (1999) has noted that the use of tau as a source of time-to-contact information is importantly limited by four factors:

1. it neglects accelerations (that is, it assumes that velocity is constant)
2. it provides information about time to contact with the eye (as opposed to objects that bypass us)
3. it requires that the object be spherically symmetrical (e.g. a round ball as opposed to a rugby ball)
4. it requires that the object's image size and rate of expansion are above perceptual threshold (i.e. it doesn't work for very small objects).

Although there is no doubt that we use time-to-contact information in controlling interceptive movements and avoiding obstacles, more recent work has established that it is only one of many such sources (Tresilian, 1999).

Vision and manual aiming

Manual aiming tasks, often called goal-directed actions, are movements that involve moving one or both hands over a prescribed distance to a target (e.g. typing on your keyboard or ringing a doorbell). It is now well established that vision can be used to correct the hand trajectory when reaching towards stationary targets and that vision is involved in different ways at different times during an aiming movement (Elliott *et al.*, 2001).

Much of the research that has examined manual aiming tasks has been concerned with the simple tasks of moving the hand from one position to another, as quickly and accurately as possible. It was this kind of task that, over 100 years ago, Robert Woodworth (1899) first studied in detail. Woodworth (1899) set up an experiment in which participants had to move a pencil back and forth through a slit at different rates specified by a metronome. Participants' movements were recorded by allowing the pencil to draw a line on a paper roll that turned beneath the work surface. In condition 1, participants were asked to make the movement with their eyes open, and in condition 2 participants were asked to close their eyes. Woodworth examined the effect of having vision on participants' accuracy (mean absolute value of the distance between the point where the pencil reversed direction and where it should have reversed direction). Data showed that when participants' eyes were closed mean absolute error remained more or less constant as velocity increased. However, when their eyes were open mean absolute error decreased as velocity decreased. Woodworth suggested that his results were due to if and when visual feedback was used to correct for errors. He believed that aiming movements comprised two components: an initial impulse or ballistic phase, and a homing-in or current control phase. In the former, participants initiate an impulse to the hand which drives it forward towards the target; he believed that this was preprogramed. During the second phase the participant uses visual information, or feedback, to correct any errors that may have occurred as the hand is moved towards the target. However, the success of this phase, according to Woodworth, depends upon whether there is sufficient time to process visual feedback (*Box 2*).

Box 2. Phases of aiming/reaching movements (Woodworth, 1899)

1. An initial-impulse phase (also called a ballistic or programed phase)
2. A current-control phase.

Keele and Posner (1968) used a discrete-task version of the Woodworth study (which used repetitive back-and-forth movements). They trained subjects to move a stylus to a small target with movement times as close as possible to 150, 250, 350 and 450 ms. On randomly selected trials, they turned the lights off when the subjects left the starting position, so that the entire movement was made in the dark (but planned in the light). Keele and Posner's (1968) data suggested that the time required to process visual feedback from these results is between 190 and 260 ms, a finding very similar to that proposed by Woodworth. Subsequently, many other studies have provided evidence that these estimates are too conservative (Paulignan *et al.*, 1991; Elliott *et al.*, 2001). The conclusion from the available evidence suggests that there is no single, *absolute* estimate of the time to process visual feedback that is likely to be correct, and the nature of the task, the performer and type of visual information are a few of many factors that will affect the time required. However, an estimate of 100–160 ms is reasonable for many tasks.

Vision and interceptive actions

Catching is, in many ways, like prehension in that it comprises two subactions: the transport of the limbs and the grasping action and this means that the human movement system is both spatially and temporally constrained. It seems reasonable to assume that vision provides advance information enabling these constraints to be met. In addition, at times when the grasping action has to be altered to retain the ball, it assumes that proprioception and tactile information are also important. Catching has attracted considerable attention as it provides a good opportunity to study the complex interactions between human motor control processes and the dynamic environment in which we live (Tresilian, 1999). Much of this research has attempted to identify factors that are involved in catching, more specifically how much visual information is required for successful interception.

Research Highlight: Lateiner and Sainburg (2003)

Lateiner and Sainburg (2003) examined the relative roles of visual and proprioceptive information about initial hand position on movement accuracy.

By employing a virtual reality environment, the experimenters were able to systematically manipulate the visual display of the hand (visual condition) start position from the actual hand (proprioceptive condition) position during movements made to a variety of directions. Subjects performed a series of baseline movements towards one of three targets in each of three blocks of trials. Interspersed among these trials were 'probe' trials in which the cursor location, but not the hand location, was displaced relative to the baseline start position. In all cases, cursor feedback was blanked at movement onset. Lateiner and Sainburg (2003) showed that subjects systematically adjusted the direction of movement in accord with the virtual location of the hand, supporting the hypothesis that visual information predominates when specifying movement direction.

Whiting *et al.* (1970) attempted to find out if there was a critical time interval for taking in flight information in ball-catching, and whether the magnitude of this period varied as a function of participant's skill level on the task. A ball was illuminated for predetermined temporal intervals of its flight path (100, 150, 200, 250, 300 and 400 ms – the latter being the total time of the flight of the ball), and subjects attempted to catch the ball one-handed. Performance was shown to be better the longer the time that was available to view the ball; however, there was little difference in the number of catches made between the 300 ms and 400 ms conditions. When Lamb and Burwitz (1988) attempted to replicate this experiment they were able to demonstrate that there was an improvement in catching performance when viewing time was increased from 100 ms to 200 ms, yet no further improvement was noted when the viewing time was increased to 300 ms or 400 ms.

Smyth and Marriott (1982) initiated a new line of inquiry with respect to catching, investigating whether sight of the hand was necessary for optimal catching performance. Smyth and Marriot believed that in skills such as catching (where vision is used to track an object) proprioception can provide the required information about limb position to ensure accurate interception, and prevention of the sight of the catching hand should, in principle, have no detrimental effect on catching performance. To test their idea Smyth and Marriot (1982) excluded vision of the hand by means of a black perspex screen attached to the side of the subject's head, which also prevented sight of the ball for the last approximately 150 ms. They reported that more position or spatial than timing or temporal errors were made when the sight of the hand was prevented. This suggested that even after a period of learning, proprioception alone does not provide accurate information for limb positioning during catching.

Fischman and Schneider (1985) used a similar paradigm to Smyth and Marriot's (1982) to examine the role of experience, in terms of skills level (i.e. experts or novices), in mediating the need to see or not to see the hand. The data collected suggest that sight of the hand is necessary for successful one-handed catching; however, more temporal errors rather than spatial errors were made during the study. The screen obscured both the hand and the ball slightly, thus could it be said that the trajectory of the ball must be seen to ensure accurate positioning and timing of the hand and thus a successful catch? The ball could be seen all the time, but for 150–200 ms prior to contact with the hand (the latency period). Fischman and Schneider propose that losing sight of the ball in such a period could be detrimental to novice catchers when positioning their hand. In combination, these results suggest that sight of the hand during catching may be important to calibrate or fine-tune the proprioceptive system, particularly with novices. Skilled catchers may have more refined articular proprioceptive receptors and subsequently are less likely to be affected by restricted visual proprioception. This delegation from vision to proprioception enables vision to be used for planning future actions or detecting environmental events. Theoretically, the assumption is that practice leads to the shift in reliance on different sources of sensory feedback or there is a transition from feedback-dependent to more open-loop control (see Section D).

Even though there are several methodological problems with research on catching, current understanding implies that visual information is helpful for the successful positioning of the hand. The relative importance of this seems to interact with the participant's skill level. It seems that accurate perception of visual information is essential in the process of learning to catch (*Fig. 5*), and as

Fig. 5. A cricketer will track the ball in an attempt to make contact at the appropriate time.

skill acquisition progresses the performer should then make more effective use of other information sources to support performance, enabling visual resources to be employed elsewhere. Land and McLeod (2000) conducted a study on cricket batting to determine where the batters were looking when facing fast bowling. Balls were delivered from a bowling machine at $25\,m\,s^{-1}$ and the following results were noted. Firstly, the fovea was directed to the ball as soon as it left the machine, at the bounce point, and for about 200 ms after the bounce. In addition, during the final part of the trajectory eyes loosely tracked the ball and large saccadic eye movements were used to get from one point to another. Interestingly they found similar patterns of visual search between performers but key difference related to ability, with the batters with the lowest skill starting saccadic eye movements later in the performance.

Researchers today are more concerned with the nature of the information, in particular what information comes from the moving object itself (e.g. Savelsbergh *et al.*, 1991). Interest in this aspect of catching stems from the theory of direct perception (Gibson, 1979). With reference to vision, Gibson (1979) emphasized that direct perception

> . . . is the activity of getting information from the ambient array of light. I call this a process of information pickup that involves the exploratory activity of looking around; getting around and looking at things. (Gibson, 1979, p.147)

With respect to catching, if we use principles of direct perception, we no longer need to conceptualize the human performer as a machine. The performer does not need to detect and use sophisticated cues from the environment in order for them to select a centrally stored representation that will enable them to catch a ball, as all the information that is needed is already available in the optic array in the form of optic variables such as tau.

Savelsbergh *et al.* (1992) had subjects catch balls projected from a distance of 6 m and at approaching speeds of 11.9, 13.9 and 16.2 m s^{-1}. For each catch, the tau-margin at the moment of initiation of the grasp was determined. It was found that muscular activity in the hand and forearm was geared to a constant tau-margin rather than the speed of the ball, thereby providing evidence for the use of tau in controlling interceptive timing. This work showed that subjects' actions were consistent with a tau-strategy; whether tau was actually used or not was put to the test by Savelsbergh *et al.* (1991). In this study, subjects were required to catch a luminous ball attached to a pendulum in a totally dark room. Three different balls were used and randomized over trials; two of the ball diameters were 5.5 cm and 7.5 cm. The third ball, however, was designed so that throughout its flight the size could be changed from 7.5 cm to 5.5 cm. This third ball specified a longer time to contact, thereby allowing a direct test of the use of tau in one-handed catching. Upon examination of the grasp, Savelsbergh *et al.* (1991) found that the moment of maximal closing velocity occurred later for the deflating ball, strongly suggesting the use of tau in the timing of the grasp.

Vision and balance

The role of vision in maintaining balance can be easily demonstrated by trying standing on one leg, much as you do when stretching the muscles at the front of your thigh, and then closing your eyes. Over 60 years ago, Edwards (1946) established that body sway is attenuated by up to 50% when the eyes are open rather than when closed, and since this time a wide variety of parameters have been manipulated to better understand the relationship between vision and balance and posture, with much of it focusing on whether the information used for balance and postural control is spatially distributed in the optic array (Lejeune *et al.*, 2006). Much of the work, particularly with adults, has shown that it is in fact peripheral vision that plays the major role in maintaining posture (Amblard and Carblanc, 1980; Stoffregen *et al.*, 1987; Nougier *et al.*, 1998; Berencsi *et al.*, 2005). However, this is not to say that central vision is not important, but the effect of manipulating this information and its effect on posture is more subtle.

Many studies have shown that vision aids proprioception in the maintenance of balance, but do not address whether or not vision can override proprioception. It was Lee and colleagues (Lishman and Lee, 1973; Lee and Lishman, 1975) who developed the 'swinging-room paradigm' to test the role of vision and whether it could override proprioception when maintaining posture. The swinging-room paradigm consisted of a large suspended box, open at the bottom and one end, which was swung around a participant who was standing on a solid and stationary floor. This experimental set-up causes the participant to experience a conflict between visual information, which suggests movement, and the proprioceptive and vestibular information available to them which suggest stability. For example, by swinging the room forward, this causes the person to think he/she is falling backwards (the visual motion of the room is consistent with this); this results in the person making a postural adjustment forwards.

Lee and colleagues showed that moving the room by 94 cm caused toddlers (mean age 15 months) to lose balance, in that they staggered, swayed or fell on 82% of trials. Thus, even though proprioceptive information told them they were in a stable upright posture, toddlers were literally 'bowled over' by the visual information (Wann *et al.*, 1998). Consistent with the age-related increase in stability and decrease in latency to regain stability (Baumberger *et al.*, 2004), Stoffregen *et al.* (1987) showed that toddlers under the age of 2 years fell more than those aged 2–5 years when the room was moved through a discrete 15 cm trajectory. In addition, there is evidence for greater postural compensations in standing infants when the walls of a moving room are moved toward them rather than away from them (Berenthal and Bai, 1989; Lejeune *et al.*, 2006).

Adult normal subjects are similarly sensitive to variations in optic flow in maintaining balance (Lee and Lishman, 1975). To compensate for what they perceived to be a loss of balance, the adults adjusted their posture, and although they swayed, it was only over a limited range. However, when information from the base of support was reduced for adults then they responded in a similar way to the toddlers (Lee and Young, 1986). Again, this phenomenon has been attributed to a conflict between the changing visual input on the one hand and the relatively constant vestibular and proprioceptive information on the other (Lestienne *et al.*, 1977).

Thus there is a great body of work that has highlighted vision's role in postural control or maintaining balance by showing that adults and children will show postural compensations to imposed optic flow, even though vestibular and somatosensory systems specifiy stasis. These compensations have been seen whether optic flow is manipulated in the swinging room paradigm (Lee and Aronson, 1974; Lee and Lishman 1975; Berenthal *et al.*, 1997) or projected onto a two-dimensional surface (Dijkstra *et al.*, 1994).

Vision and locomotion

The visual system is uniquely positioned to provide information on the static and dynamic features of our environment, which we can use to locomote. A common locomotor problem that we all are confronted with is the need to negotiate an object or obstacle. Although kinesthetic and vestibular systems play an important role when walking, particularly when reacting to perturbations, it is preferable to avoid these perturbations, which is made possible by using visual information available from our environment. *Table 1* provides an overview of receptors and the type of information they provide. Perturbation experiments

Table 1. The type of information produced by each receptor organ and its transducer

Receptor organ	Transducer	Type of information
Eyes	Light detector	Vision
Ears	Sound detector	Hearing
Vestibular system (e.g. utricle, saccule)	Linear and angular accelerations	Balance
Cutaneous receptors	Pressure, stretch deformation	Touch, texture, pressure
Muscle spindles	Length and velocity	Muscle length and velocity, joint position
Golgi tendon organs	Force and tension	Force
Joint receptors	Mechanical deformation	Joint movement
Nocioceptors	Mechanical impacts	Pain, impact

have shown that visual information can be used to alter foot trajectory rapidly during a step; for example after presentation of an obstacle, research has shown that changes in foot trajectory are seen within 120 ms of the visual stimulus (Patla *et al.*, 1991; Reynolds and Day, 2005).

The visual information we refer to is contained in patterns of optic flow that result from the dynamic projection of the visual scene onto the retina during an observer's displacement (Gibson, 1979). A classic experiment demonstrating the influence of time-to-contact information on locomotion was carried out by Lee *et al.* (1982) on the approach of long jumpers to the take-off board. They examined the stride length changes of three highly skilled female long jumpers for a series of six long jumps. For the first five or six strides the athletes' stride length increased at a relatively constant rate; it then stayed relatively constant for the next six strides, and then on the final six strides, the athletes made stride-length adjustments so that they could hit the board accurately. The authors suggested that these data indicate that the step pattern during the run-up to each jump is not programed entirely in advance. In fact, Lee *et al.* (1982) argued that the approach phase of the long jump consisted of two phases: an accelerative phase and a zeroing-in phase. It was suggested that although the athletes tried to make each run with a fixed number of steps, the step lengths were not held constant and these inconsistencies resulted in an unavoidable cumulative effect which resulted in the build-up of footfall variability until the fifth-from-last step (Scott *et al.*, 1997). Thus the stride-length adjustments during the last few steps correct for the distance error that has accumulated during the run-up. Lee *et al.* (1982) concluded that the zeroing-in phase was achieved by visual regulation of stride length. More specifically, the authors suggested that the optic variable 'tau' was coupled to the impulse imparted by the long jumper on the take-off board. Overall Lee *et al.* (1982) suggest that the long jumper bases the correction process on visual information obtained in advance of these strides, i.e. time-to-contact information obtained visually from the board, rather than distance to contact.

The findings of Lee *et al.* (1982), for the pattern of variability in footfall placement during the approach phase, have been replicated with larger samples of elite subjects (Hay, 1988) and novice athletes (Scott *et al.*, 1997). Scott *et al.* (1997) examined the consistency of foot placement over trials on 11 non-long jumpers. In comparison to more skilled subjects, non-long jumpers accumulated a considerably larger maximum mean standard deviation in footfall placement between trials. However, the results also showed that non-long jumpers have a similar pattern of descending variability near to the take-off board to that of expert long jumpers, suggesting the use of visual regulation.

Bradshaw (2004) utilized the standard deviation method of Lee *et al.* (1982) for identifying visual regulation in long jump run-ups, to examine whether visual regulation processes are used in the run-up of gymnastic vaulting. Five elite female gymnasts aged 13–15 years performed five round-off entry vaults while being filmed with digital cameras. Two qualified judges viewed each vaulting trial and provided a performance score. The data showed that a precursor for a fast take-off from the board when vaulting is to utilize vision early to control the approach kinematics, and that high take-off velocity was directly related to judge's score. This suggested that visual regulation of the stride pattern is evident and that this leads to a larger impulse at take-off and a better score.

Conclusion

Sensory information is vital to control, and the human system is able to use sensory information in a variety of ways to control and refine movement. As we move about the environment we are constantly making adjustments; many of these adjustments are so fine that we are unaware that we are making them. This section has outlined the role of sensory information for control and coordination and also covered a range of fascinating research which, over the years, has revealed how we are able to do so in complex and diverse movement contexts.

Further reading

Land, M.F. and Mcleod, P. (2000). From eye movements to actions: how batsman hit the ball. *Nat Neurosci* **3**, 1340–1345.

Myers, J.B., Craig, A. Wassinger, C.A., Scott, M. and Lephart, S.M. (2006) Sensorimotor contribution to shoulder stability: effect of injury and rehabilitation. *Man Ther* **11**, 197–201.

Vuillerme, N., Teasdale, N. and Nougier, V. (2001) The effect of expertise in gymnastics on proprioceptive sensory integration in human subjects. *Neurosci Lett* **311**, 73–76.

Witney, A.G., Wing, A., Thonnard, J.L. and Smith, A.M. (2004) The cutaneous contribution to adaptive precision grip. *Trends Neurosci* **27**, 637–643

References

Amblard, B. and Carblanc, A. (1980) Role of foveal and peripheral visual information in maintenance of postural equilibrium in man. *Percept Mot Skills* **51**, 903–912.

Baumberger, B., Isableu, B. and Flückiger, M. (2004) The visual control of stability in children and adults: postural readjustments in a ground optical flow. *Brain Res Exp Brain Res* **159**, 33–46.

Berencsi, A., Ishihara, M. and Imanaka, K. (2005) The functional role of central and peripheral vision in the control of posture. *Hum Mov Sci* **24**, 698–709.

Berenthal, B. and Bai, D.L. (1989) Infants' sensitivity to optic flow for controlling posture. *Dev Psychol* **25**, 936–945.

Berenthal, B.I., Rose, J.L. and Bai, D.L. (1997) Perception-action coupling in the development of visual control of posture. *J Exp Psychol Hum Percept Perform* **23**, 1631–1643.

Bizzi, E. and Polit, A. (1979) Processes controlling visually evoked movements. *Neuropsychologia* **17**, 203–213.

Bossom, J. (1974) Movement without proprioception. *Brain Res* **71**, 285–296.

Bradshaw, E. (2004) Target-directed running in gymnastics: a preliminary exploration of vaulting. *Sports Biomech* **3**, 125–144.

Dijkstra, T.M.H., Schöner, G. and Gielen, C.C.A.M. (1994) Temporal stability of the action perspective cycle for postural control in a moving environment. *Brain Res Exp Brain Res* **97**, 477–486.

Edwards, A. (1946) Body sway and vision. *J Exp Psychol* **36**, 526–535.

Elliott, D., Helsen, W.F. and Chua, R. (2001) A century later: Woodworth's (1899) two-component model of goal-directed aiming. *Psychol Bull* **127**, 342–357.

Fischman, M.G. and Schneider, T. (1985) Skill level, vision, and proprioception in simple one-hand catching. *J Mot Behav* **17**, 219–229.

Fuchs, S.M.D., Thorwesten, L. and Niewerth, S. (1999) Proprioceptive function in knees with and without total knee arthroplasty. *Am J Phys Med Rehabil* **78**, 39–45.

Gibson, J.J. (1966) *The Senses Considered as Perceptual Systems.* Houghton Mifflin Company, Boston.

Gibson, J.J. (1979) *The Ecological Approach to Visual Perception.* Houghton-Mifflin Company, Boston.

Hay, J.G. (1988) Approach strategies in the long jump. *Int J Sport Biomech* **4**, 114–129.

Hay, L., Bard, C., Ferrel, C., Olivier, I. and Fleury, M. (2005) Role of proprioceptive information in movement programming and control in 5 to 11-year old children. *Hum Mov Sci* **24**, 139–154.

Hay, L., Bard, C., Fleury, M. and Teasdale, N. (1996) Availability of visual and proprioceptive afferent messages and postural control in elderly adults. *Brain Res Exp Brain Res* **108**, 129–139.

Keele, S.W. and Posner, M.I. (1968) Processing of visual feedback in rapid movements. *J Exp Psychol* **77**, 155–158.

Knapp, H., Taub, D.E. and Berman, A.J. (1963) Movements in monkeys with deafferented forelimbs. *Exp Neurol* **7**, 305–315.

Lamb, K.L. and Burwitz, L. (1988) Visual restriction in ball-catching: a re-examination of earlier findings. *J Exp Psychol* **77**, 155–158.

Land, M.F. and Mcleod, P. (2000) From eye movements to actions: how batsman hit the ball. *Nat Neurosci* **3**, 1340–1345.

Lappe, M., Bremmer, F. and Van den Berg, A.V. (1999) Perception of self-motion from visual flow. *Trends Cogn Sci* **3**, 329–336.

Lateiner, J.E. and Sainburg, R.L. (2003) Differential contributions of vision and proprioception to movement accuracy. *Brain Res Exp Brain Res* **151**, 446–454.

Lee, D.N. (1976) A theory of visual control of braking base on information about time-to-collision. *Perception* **5**, 437–439.

Lee, D.N. and Aronson, E. (1974) Visual proprioceptive control of standing in human infants. *Percept Psychophys* **15**, 529–532.

Lee, D.N. and Lishman, J.R. (1975) Visual proprioceptive control of stance. *J Hum Mov Stud* **1**, 87–95.

Lee, D.N., Lishman, J.R. and Thomson, J.A. (1982) Regulation of gait in long jumping. *J Exp Psychol Hum Percept Perform* **8**, 448–459.

Lee, D.N. and Young, D.S. (1986) Visual timing of interceptive actions. In: Ingle, D.J., Jeanerod, M. and Lee, D.N. (eds) *Brain Mechanisms and Spatial Vision*. Martinus Nijhoff, Dordrecht, pp.1–30.

Lejeune, L., Anderson, D.I., Campos, J.J. *et al*. (2006) Responsiveness to terrestrial optic flow in infancy: does locomotor experience play a role? *Hum Mov Sci* **25**, 4–17.

Lestienne, F., Soeching, J. and Berthoz, A. (1977) Postural readjustments induced by linear motion of visual scenes. *Brain Res Exp Brain Res* **28**, 363–384.

Lishman, J.R. and Lee, D.N. (1973). The autonomy of visual kinaesthesis. *Perception* **2**, 287–294.

Loucks, T.M.J. and De Nil, L.F. (2001) The effects of masseter tendon vibration on nonspeech oral movements and vowel gestures. *J Speech Lang Hear Res* **44**, 306–316.

Mazzaro, N., Grey, M.J., Sinkjaer, T. *et al*. (2005) Lack of on-going adaptations in the soleus muscle activity during walking in patients affected by large-fiber neuropathy. *J Neurophysiol* **93**, 3075–3085.

Nougier, V., Bard, C., Fleury, M. and Teasdale, N. (1998) Contribution of central and peripheral vision to the regulation of stance: developmental aspects. *J Exp Child Psychol* **68**, 202–215.

Patla, A.E (1997) Understanding the roles of vision in the control of human locomotion. *Gait Posture* **5**, 54–69.

Patla, A.E., Beuter, A. and Prentice, S. (1991) A two stage correction of limb trajectory to avoid obstacles during stepping. *Neurosci Res Commun* **8**, 153–159.

Paulignan, Y., MacKenzie, C., Marteniuk, R. and Jeannerod, M. (1991) Selective perturbation of visual input during prehension movements. 1. The effects of changing object position. *Brain Res Exp Brain Res* **83**, 502–512.

Polit, A. and Bizzi, E. (1978) Processes controlling arm movements. *Science* **201**, 1235–1237.

Quoniam, C., Hay, L., Roll, J. P. and Harlay, F. (1995) Age effects on reflex and postural responses to propriomuscular inputs generated by tendon vibration. *J Gerontol A Biol Sci Med Sci* **50**, B155–B165.

Reeve, T.G., Mackey, L.J. and Fober, G.W. (1986) Visual dominance in the cross-modal kinesthetic to kinesthetic plus visual feedback condition. *Percept Mot Skills* **62**, 243.

Reynolds, R.F. and Day, B.L. (2005) Rapid visuo-motor processes drive the leg regardless of balance constraints. *Curr Biol* **15**, R48–R49.

Savelsbergh, G.J.P., Whiting, H.T.A. and Bootsma, R.J. (1991) Grasping tau. *J Exp Psychol Hum Percept Perform* **17**, 315–322.

Savelsbergh, G.J.P., Whiting, H.T.A., Burden, A.M. and Bartlett, R.M. (1992). The role of predictive visual temporal information in the coordination of muscle activity in catching. *Brain Res Exp Brain Res* **89**, 223–228.

Scott, M.A., Li, F.X. and Davids, K. (1997) Expertise and the regulation of gait in the approach phase of the long jump. *J Sport Sci* **15**, 597–605.

Simoneau, G.G., Degner, R.M., Kramper, C.A. and Kittleson, K.H. (1997) Changes in ankle joint proprioception resulting from strips of athletic tape applied over the skin. *J Athl Train* **32**, 141–147.

Smyth, M.M. and Marriott, A.M. (1982) Vision and proprioception in simple catching. *J Mot Behav* **14**, 143–152.

Stoffregen, T.A., Schmuckler, M.A. and Gibson, E.J. (1987) Use of central and peripheral optical flow in stance and locomotion in young walkers. *Perception* **16**, 113–119.

Taub, E. and Berman, A.J. (1963) Avoidance conditioning in the absence of relevant proprioceptive and exteroceptive feedback. *J Comp Physiol Psychol* **56**, 1012–1016.

Taub, E., Heitmann, R.D. and Barro, G. (1977) Alertness, level of activity, and purposive movement following somatosensory deafferentation in monkeys. *Ann NY Acad Sci* **290**, 348–365.

Tresilian, J.R. (1999) Analysis of recent empirical challenges to an account of interceptive timing. *Percept Psychophys* **61**, 515–528.

Wann, J.P., Mon-Williams, M. and Rushton, K. (1998) Postural control and coordination disorders: the swinging room revisited. *Hum Mov Sci* **17**, 491–513.

Whiting, H.T.A., Gill, E.B. and Stephenson, J.M. (1970) Critical time intervals for taking in flight information in a ball-catching task. *Ergonomics* **13**, 265–272.

Woodworth, R.S. (1899) The accuracy of voluntary movements. *Psychol Rev Monogr Suppl* **3**(3).

THEORIES OF MOTOR LEARNING

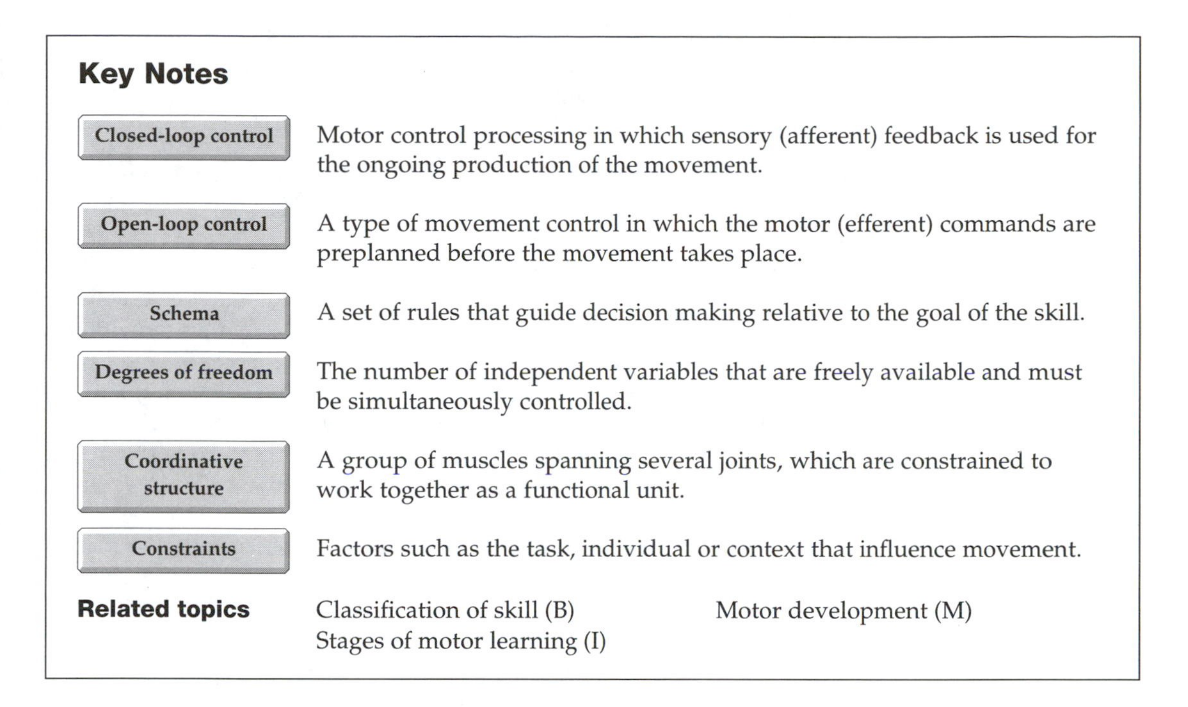

Key Notes

Closed-loop control	Motor control processing in which sensory (afferent) feedback is used for the ongoing production of the movement.
Open-loop control	A type of movement control in which the motor (efferent) commands are preplanned before the movement takes place.
Schema	A set of rules that guide decision making relative to the goal of the skill.
Degrees of freedom	The number of independent variables that are freely available and must be simultaneously controlled.
Coordinative structure	A group of muscles spanning several joints, which are constrained to work together as a functional unit.
Constraints	Factors such as the task, individual or context that influence movement.

Related topics Classification of skill (B) Motor development (M) Stages of motor learning (I)

Any individual involved in sport or rehabilitation should have a solid understanding of theories of motor learning and stages of learning, and be able to assess this process. Motor learning is concerned with understanding the relatively permanent changes in the capability to produce skilled action that is acquired with practice or as a result of experience. When watching a game of football you have to wonder how premier league players are capable of dribbling the ball down the field and then passing the ball with such precision, all without paying very little direct attention to the mechanics of their performance. Their movements are in stark contrast to those of the novice who watches their feet in order to keep control of the ball and has difficulty performing the fundamental kicking skills needed to deliver the ball to their team mates. The actions of a novice usually require a considerable amount of physical and mental effort as they attempt to discover a solution to the movement problem. In this section we will consider motor learning by firstly considering theories of motor learning and then we will address stages of learning that account for the progression from novice to expert. In addition, this section will also examine motor memory, which is a fundamental component of learning.

Theories of motor learning

While the focus of motor control is understanding the control of movement already acquired (see Section D), the field of study concerned with the description and explanation of how we acquire and/or modify movements that accompany practice is typically referred to as motor learning.

Two terms that need to be defined (see Section C) are motor learning and performance (*Box 1*).

Box 1. Definitions

- *Motor learning*: 'a change in the capability of a person to perform a skill that must be inferred from a relatively permanent improvement in performance as a result of practice or experience'. (Magill, 2007, p.247)
- *Performance*: the observable behavior during the execution of a skill at a specific time in a specific context.

According to many researchers (Schmidt *et al.*, 1979; Schmidt and Young, 1987; McDonald *et al.*, 1989; Newell and Vaillancourt 2001), motor learning is described as a set of interrelated processes that are associated with experience or practice, leading to relatively permanent changes in the capability to produce action. At the same time, it is acknowledged that learning cannot be measured directly, but needs to be inferred from observable changes in an individual's performance. However, a more contemporary view is that motor learning is far more than a set of processes; moreover it involves an active search for, and the subsequent modification of, many movement patterns to solve a movement problem (*Fig. 1*). Such task solutions are new strategies for perceiving and acting which emerge due to constraints imposed by the individual, task and environment (Newell, 1991).

There is no doubt that thousands of hours of practice separate the performance of the expert and novice (Ericcson *et al.*, 1993). However, given the huge number of solutions to every movement problem, theory-driven explanations of how this practice should be implemented in different stages of the learning process can help us better understand how we can foster the learning of many motor skills. It will become clearer throughout this book that it is not only the amount of practice that is related to skill level and learning, but that

Fig. 1. Performance is influenced by the level of skill of the player, the rules of the game, and the environmental conditions.

combination of how the task is practiced and how the environment is manipulated depending upon the skill level of the performer, which eventually determines how well a skill is learned.

Motor-learning theories that guide teachers, coaches and therapists today have emerged gradually as either a compilation of, or reaction to, older theories of motor learning. Adams' (1971) closed-loop theory is recognized as the first comprehensive explanation of motor learning. Adams stated that although the central nervous system (CNS) controlled the execution of movement, it had to be based upon sensory feedback. The acknowledgment of the performer using extrinsic feedback was an important step in understanding how we learn movements. Throughout the 1970s, hierarchical models continued to dominate the understanding of motor learning and skill development, with Schmidt's schema theory (1975) forming the basis of many of the proposed hierarchical theories of motor learning (see Newell *et al.*, 2003).

Motor-learning theorists gradually began to reject the concept of a centralized, hierarchically organized set of motor programs. With a strong emergence of an ecological approach to perception and action which incorporated both Gibson's ecological theory of direct perception (see Section D) and Bernstein's (1967) concepts of movement coordination, theorists began to argue that movement skill is an emergent property of the interaction of multiple, cooperative systems (Thelen *et al.*, 1987) and the acceptance of the idea that it is the interactions between the performer, action goal, and the features of the environment that are critical to understanding principles of motor learning today.

The next part of this section will examine the basic foundations of each of these theories as they are related to the acquisition of motor skill within the area of motor control.

Adams' closed-loop theory

Adams (1971) believed that all motor acts are dependent on a matching process between the desired movement, stored in movement memory, and the actual movement being produced. Thus, memory serves an active role, guiding the current movement by supplying information to the performer in the form of ongoing feedback from limbs. When an individual makes a positioning movement such as reaching for a glass, inherent feedback is produced that represents the particular location of the limb in space and whether or not the movement was effective (inherent feedback is also referred to as intrinsic feedback). These stimuli 'leave a trace' in the CNS – the perceptual trace. The perceptual trace is a composite of all the sources of information produced by the action: proprioceptive and tactile inputs that arise from the displacement of the limbs, as well as visual and auditory information that are a result of the movement. With practice over a number of trials, the individual will come closer and closer to the target or, in the case of reaching for a glass, each hand position along a trajectory will be perceptually defined making the glass grabbing easier as trace after trace is laid down. These collections of traces come to represent the feedback qualities of the correct movement. As the performer receives feedback, the perceptual trace becomes stronger, hence the errors in performance decrease with practice until the perceptual trace comes to provide an accurate reflection of the skilled movement. Thus in any movement:

- *the memory trace* selects and initiates a plan of action
- *the perceptual trace* compares the movement in progress with a correct memory of the movement.

Adams (1971) argues that learning reflects the development of more adaptive perceptual traces as well as more adaptive capacities for generating movements that reduce errors between perceptual traces and the outcomes (*Fig. 2*).

A central tenet of the closed-loop theory of learning is the explicit need for feedback in terms of knowledge of results (KR). Adams states that KR provides information to solve the motor problem; after each trial the feedback given provides information about how the next movement should be made better to achieve the task goal. Thus, in the early stages of acquisition of a skill KR is very important. As more correct movements are produced, the perceptual trace is solidified and KR is no longer needed. One of the predictions of the theory is that learning is still possible even when KR is eliminated. Indeed, in the late stages of learning the performer has very little use of KR. The above prediction has received support from a number of studies (Adams *et al.*, 1972; Newell, 1974), despite the prediction being very difficult to test as the small changes achievable at this stage may be obscured by measurement noise (Ivry, 1996).

Contrary to earlier closed-loop theorists, Adams realized that in order for the system to have the capacity to detect its own errors, two memory states must be present. The perceptual trace represents the correct movement, and the movement is selected and initiated by the memory trace. The perceptual trace then takes over control again to guide the movement to its final target location. The memory trace requires neither feedback nor a requisite context in order for a response to be selected. Selection of a memory trace will lead to the onset of action, at which point the skilled execution of the movement will be carried out

Fig. 2. In the early stages of acquisition of a skill, knowledge of results is very important.

under the guidance of the error-sensing perceptual trace. If either the memory trace or perceptual trace is diminished, learning is less than optimal.

The two-component aspect of Adams' theory was tested by Newell and Chew (1974) by equating the perceptual and memory trace with recall and recognition memory traces. Newell and Chew observed dissociation between these two processes, and when KR was withdrawn performance (recall) did not deteriorate, but the subjects' estimates of their performance was more erroneous (recognition error). Newell and Chew (1974) viewed this dissociation as evidence to support the proposition that the memory and the perceptual traces are independent.

Rosenbaum (1991) documents problems associated with the closed-loop theory, stating the theory can be ultimately reduced to a response-chaining theory; each movement, or component of a movement is assumed to be triggered by a perceived error, yet many movement sequences can be performed effectively when feedback is removed. Indeed, even complex movements can often be performed effectively without proprioceptive, visual or other forms of feedback. The theory also incorrectly predicts that the more KR a learner receives the more effectively he/she will perform (Winstein and Schmidt, 1990; Wulf and Schmidt, 1996; Weeks and Kordus, 1998) (see Section L, p.171). Indeed a number of disadvantages of Adams' theory of learning have been identified; these are listed in *Box 2*.

Box 2. Advantages and disadvantages of Adams' (1971) theory of learning

Advantages

- Makes predictions about learning which can be tested
- Concerned with both learning and performance
- Built on sound theoretical issues, many of which are testable.

Disadvantages

- Adams' work has been concerned solely with *slow positioning movements*. It has inherent problems with tasks involving fast movement, or of a ballistic nature.
- *Problems of storage*: Adams is arguing for a perceptual trace for every movement but we do not know how this is represented in the brain. When you consider the range of movements we can perform, a perceptual trace for each of these would put tremendous strain on any memory system.
- It fails to account for accurate *novel movements*. We can perform novel movements accurately. Furthermore, it is unlikely that in everyday life we perform the same movement twice, even in reproducing the same end result.
- *In the absence of KR, the difference between the perceptual trace and the actual response, in a slow positioning task provides no KR*, and therefore no comparison. There is no possibility of comparing the perceptual trace as the memory trace only exists in terms of the perceptual trace which was executed. This cannot lead to learning without KR, as learning dictates analysis of the memory trace. It will however, perhaps lead to a great consistency in performance, even if consistently inaccurate. Each response will be, or attempt to be, approximated to the previous perceptual trace, in the absence of a memory trace.

To account for the competence of movements made in the absence of feedback, theorists have postulated open-loop models of motor control, central to which is the concept of the motor program. Schmidt acknowledged that the sensory information that Adams believed was used to account for control of actions was too slow to be used when fast actions are produced. Therefore, Schmidt proposed that fast actions had to be organized in advance and were represented in the memory as motor programs.

Schmidt's schema theory

Schmidt (1975) put forth a comprehensive learning theory that embraces the notion of a motor program in a generalized form. The unique feature of Schmidt's theory was the generalized motor program (GMP). This motor program did not contain instructions detailing the specifics of the movement, but rather generalized instructions for a given class of movements; this could thus account for the number of variations with which an action could be performed. For example jogging, skipping, and running are all the same class of movement. The GMP can be scaled in time (by scaling a rate parameter) and amplitude (by scaling a force parameter), thus relative timing and force (i.e. the specification of) are rigidly structured 'in' the program (Schmidt, 2003). Thus the performer has to learn two things from practice:

1. acquire a GMP or defined form
2. learn schemata that allow the action to be scaled to the environment.

Schmidt's (1975) schema theory of motor control considered the criticisms that were made of Adams' theory, and attempted to account for them. Schmidt was especially concerned with discrete tasks, of both the slow positioning type and the ballistic type. In addition he looked at both open and closed skills, and he also considered skills that demanded high accuracy and those that have as their goal maximum speed, height, etc. The notion of a schema is not a new idea; Head (1926) made the first formal description of a schema. Bartlett (1932) subsequently modified the idea, so it has been around for at least the last 70 years. *The Dictionary of the Sport and Exercise Sciences* gives a definition that is related to sport:

> Set of rules, formed by abstracting information from related movement experiences, that includes the underlying movement principle for this class of responses. (*Dictionary of the Sport and Exercise Sciences*, 1991, p.133)

Bartlett (1932) was the first to try and relate the idea of a schema to motor responses. He argued that in making a tennis stroke, the individual depends on relating new experiences (mainly visual) to other immediately preceding experiences, such as posture and balance, at that moment in time. How this actually occurred was not suggested by Bartlett, and it was not until the 1970s that this concept began to be studied.

Schmidt took the concept of a schema and related it specifically to motor learning. He viewed the schema as a relatively stable and permanent construct, although to a certain extent dynamic in that it is subject to modification, which is representative of a certain class of event abstractions or rules about those events. In all contexts Schmidt considered that learning was closely linked with memory and he proposes two stages:

1. *recognition*: evaluation of the response, and any corrections possible, to the ongoing movement
2. *recall*: responsible for movement production.

Thus he overcame the problems faced by Adams' theory where the same mechanism is responsible for both processes (Adams has two traces, perceptual and memory, but they are within the same mechanism). Schmidt assumed that a motor program exists somewhere within the CNS and that it probably exists in a generalized version. The motor program is at various times dependent on the two states of memory (recognition and recall). The role of these states will depend on the duration and type of movement. In slow movement, such as the linear positioning task, recognition will play an important part in the correction of error as an ongoing process. However, where the response is fast, with no opportunity for ongoing control, recall takes control, with recognition only being of value at the termination of the response.

However, whenever a movement is made, four pieces of information are made available, and it is this information that forms the basis of recognition and recall.

These four pieces of information are:

1. *initial conditions*: information received prior to the response, such as the proprioceptive information about the position of the limbs and body, and visual information about the state of the environment
2. *response specifications*: the speed, force, degree, etc of movement, which serves as a record after the execution of the movement
3. *sensory consequences*: this is information stored after the movement, which consists of the actual feedback stimuli received from the eyes, ears, proprioceptors, etc, i.e. an exact copy of the afferent information provided on the response
4. *response outcome*: the success of the response in relation to the outcome intended. It is the actual outcome that is stored, not the original goal. A subject without feedback does not have outcome information to store.

In order to formulate a schema, after several experiences of a similar kind, the individual begins to extract information from the relationships between these four elements. It is, then, these relationships that form the schema, not the information itself. The strength of the relationship between the four stored elements increases with each successive movement of the same general type, and also with increased accuracy of feedback information from the response outcome. This results in the formation of two schemas: recall schema and recognition schema.

Recall schema

On the basis of relationships between past experiences and response specifications (the recall schema), the performer determines a set of response specifications that are considered will best achieve the desired outcome. The important part is that the performer need never have executed this response or movement before, i.e. it maybe novel.

Recognition schema

As the subject is generating the response specification, they also generate a set of expected sensory consequences. These expected sensory consequences are compared with the actual sensory consequences and the error fed back into the schema. The expected sensory consequences are dependent on the relationship between actual outcomes and sensory consequences of past experiences, and this in itself forms the basis of the recognition schema.

The GMP was certainly one of the most important parts of the schema theory, and was also of interest in movement control (Schmidt, 2003). The GMP concept itself has received considerable scrutiny over the years, and three of the more widely investigated aspects of schema theory are the GMP and parameter distinction, invariant relative timing and force, and the variability-of-practice hypothesis.

Variability-of-practice hypothesis

The GMP theory predicts that variable practice can lead to better transfer to new tasks than consistent practice (Schmidt, 1975; Shapiro and Schmidt, 1982) and will prove most effective in the long-term development of a stable schema. The simplicity of the variability-of-practice hypothesis has attracted a great deal of research attention (Handford *et al.*, 1997), resulting in early widespread claims for strong empirical support (McCracken and Stelmach, 1977; Kelso and Norman, 1978; Del Rey *et al.*, 1982; Lee *et al.*, 1985). However, an extensive review of the literature by Van Rossum (1990) cast considerable doubt on the consistency of support from investigations conducted over the last 15 years. More recent published research is beginning to emerge that questions whether high interference produces better results on transfer and retention tasks indicating a higher degree of learning has taken place and hence that the schema is more stable. This key concept will be discussed in more detail on p.161, and the key aspects of conducting learning studies are discussed in Section C.

Research Highlight: Wrisberg and Lui (1991)

Using badminton tasks, Wrisberg and Lui (1991) conducted an experiment in a standard physical education instructional setting to determine the effect of contextual variety on learning and transfer of the short and long serve. Although little difference was observed in the performance among the blocked and varied groups during practice, the varied group showed higher retention of the short serve and higher transfer of both serves. In contrast, French *et al.*, (1990) obtained no significant differences in acquisition and retention for three groups of high school children, assigned to three practice conditions (two random-blocked and one blocked practice) wherein the students practiced the volleyball forearm pass, set and overhead serve. Shortly after, Bortoli *et al.* (1992) utilized French *et al.*'s (1990) method with students practicing the volleyball skills of the bump, volley and the serve. High contextual interference, even though causing immediate limited performance, led to superior performance on the retention and transfer tasks, particularly when performing the serve; thus the authors claimed that the results are partially in line with those generally found in laboratory experiments.

The generalized motor program and parameter distinction

According to the schema theory, a GMP is developed over practice and becomes a basis for generating responses that share the same invariant features within a movement class, with specific movements being produced by the premovement assignment of scaling factors, termed parameters via recall schema. A GMP is a program specifying the sequence among muscles or relative timing and relative forces among the muscle contractions (structural features). For a given GMP, the sequence of relative timing and forces are assumed to be essentially invariant even when superficial features such as overall movement duration and movement size are altered. These latter aspects (often called metrical features) are thought to be controlled by parameters that adjust the surface expression of the GMP output yet allow it to retain its invariant structure or pattern. This GMP model allows the same pattern of activity with different speeds, sizes or limbs; however, relative timing and forces are invariant.

In recent years, a number of researchers supported the GMP parameter distinction by providing evidence for an empirical dissociation of these theoretical constructs. For example, Wulf *et al.* (1993) showed it was possible to decompose errors in various movement patterning tasks into measures related to (a) errors in the scaling of the fundamental pattern (the GMP) in space and time; and (b) errors in the fundamental pattern itself. More interestingly work by Schmidt and colleagues (Wulf and Schmidt, 1989; Wulf *et al.*, 1993, 1994) showed that factors affecting one kind of measure (e.g. the GMP) do not affect learning scaling (parameterization). For example, reducing the frequency of KR during serial practice facilitates GMP development (Wulf *et al.*, 1993) but degrades the learning of dependent variables, including parameterization (Wulf and Schmidt, 1996). Lai and Shea (1999), after replicating the findings of Wulf *et al.* (1993), proposed that acquisition factors (feedback, practice manipulations) that promote trial-to-trial response stability (e.g. decreasing frequency of KR, bandwidth KR or constant practice) enhance GMP development. Conversely, acquisition factors promoting trial-to-trial parameter variability appear to facilitate the learner's capacity to parameterize precisely a movement under transfer conditions (Sherwood and Lee, 2003). As you will see from Further reading, researchers are still interested in testing and evaluating schema theory and the GMP.

Invariant relative timing and force

Many studies have examined the idea that relative time (the temporal structure) was invariant and that some 'rate parameter' could scale this structure proportionately in time. Recently, the evaluation of relative timing invariance has been extended to bimanual actions. Heuer *et al.* (1995) showed a very strong temporal invariance when participants were asked to produce simultaneous but different rapid actions with the two arms; they demonstrated that the temporal structure of each limb, as well as the temporal structure across limbs, was scaled linearly in time – that is, invariant relative timing was present for the whole task.

Heuer (1991) argued that the tendency toward temporal invariance is compelling enough that the minor deviations that have been reported should be ignored, although it must be acknowledged that there are many kinds of actions for which temporal invariance does not hold. Many of these tasks are such that the performer cannot plan the action in advance because of the need to adjust to environmental disturbances and response-reproduced variability. Goal-directed reaching is an example of this, for which the initial parts of the action appear to be scaled proportionately in time, but when the end of the action (including the grasp) is included; the whole reach-to-grasp action does not show temporal invariance. It is as if this latter phase occurs as a separate action (Martenuik *et al.*, 1990). Clearly, there are limits to the assumption that all discrete movements are run with a GMP, and perhaps a newer theory must recognize this concern.

One advantage of such a broad model of motor learning is that it recognizes that a diverse set of information-processing mechanisms takes place during the course of skilled action, and unlike the closed-loop model proposed by Adams, the model can be applied beyond simple, discrete movements to complex skills. Schmidt's (1975) schema theory has brought the area of motor control into the mainstream of cognitive psychology in terms of a conceptual explanation of learning in which it is viewed not as a process in which every event is uniquely stored, but that the schema provides the essential features adding specific characteristics as a function of the context in which the action emerges (Ivry, 1996). However, Schmidt's theory has been attacked on a number of grounds, and the advantages and disadvantages can be seen in *Box 3*.

Box 3. Advantages and disadvantages of Schmidt's (1975) theory of learning

Advantages

- Accounts for more types of movement
- Accounts for error detection more effectively
- Explains production of novel responses.

Disadvantages

- It cannot explain the development of new patterns of coordination
- It is unable to account for spontaneous compensations to perturbations or changes that occur in the environment during a movement
- It overemphasizes the role of cognitive processes
- Evidence does not always support theoretical predictions.

In Schmidt's theory, the human performer is likened to a machine that receives sensory input that it acts upon to produce an output. More specifically, it is suggested that performers detect and use sophisticated cues from the environment to select a centrally stored representation that will solve the motor problem. However, a number of influential theorists have attacked this stance and have supported their arguments with principles and concepts of dynamical systems theory, ecological psychology and a constraint-led approach to the development of coordination. All of these approaches to movement coordination highlight the importance of the interaction between the performer and the dynamics of the environment in which she or he is moving.

Ecological theory

In human performance, the idea of a one-to-one mapping between the movement commands and the action patterns does not concur with the variety of ways in which the performer achieves their goals. Indeed, there is little evidence to suggest that any movement can be reproduced in a stereotyped manner, and thus the notions of a central command structure seem unconvincing. When a performer undertakes an action, physical and biological conditions change and require subsequent changes to an established pattern of neural signal for the completion of action. This phenomenon of movement performance has been termed context-conditioned variability (Turvey *et al.*, 1982). Often the same actions may be achieved by different muscle groups, and different actions can result from the same movement patterns given a particular context of forces (Turvey *et al.*, 1982); thus the idea of a prior preparation of functional actions fails to take into account the ongoing energy exchanges that occur between the performer and his/her environment.

Central to the idea of an ecological account of motor learning is the idea that the learner seeks to discover lawful properties of the environment that make it possible to learn skills. Put another way, the performer increases the coordination between perception and action consistent with the task and environmental constraints, while controlling the huge number of degrees of freedom available at any one time (*Fig. 3*).

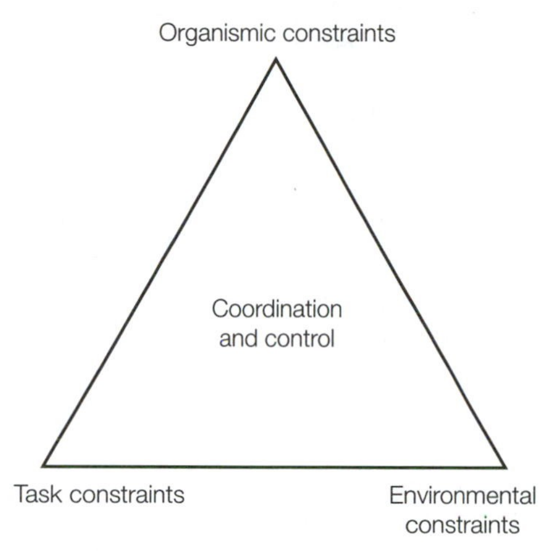

Fig. 3. Constraints that interact and guide the system. Adapted from Newell (1986).

Constraints theory (Newell, 1986)

Many researchers believe that a dynamical systems approach has much to offer when considering skill development or acquisition. A fundamental issue is the stable organization of the movement system and the processes by which this changes over time (Williams *et al.*, 1999). It must be remembered that Bernstein (1967) considered the major factor in learning to be how we master redundant degrees of freedom in order to achieve coordinated movement. This is a question that is still being explored by many researchers interested in learning and development.

Newell (1986) suggested that the process of motor learning is concerned with discovering optimal motor solutions through practices that are in line with the performer's own capabilities, the type of task the performer is learning and the environmental context it is performed in. Newell proposed that there are three major constraints that interact during the performance of coordinated movement. These are *organismic constraints, task constraints* and *environmental constraints* (*Fig. 3*).

- *Organismic constraints*: this refers to the resources that each individual has that can impact the performance of a task. These can be both strengths and weaknesses. In the sporting context, an individual arrives with a certain level of practice or past experience; they may be very strong, may lack flexibility, or may have a movement problem. All of these factors are unique to the individual and influence how they are able to perform the task. It is also important to remember that each individual has a number of subsystems (cognition, perception etc) which develop at different rates, are interconnected, and therefore have the potential to act as an organismic constraint (Thelen, 1995).

- *Task constraints*: the specific components of the task such as the rules, the equipment to be used, the tactics, and the end goal of the action are all task constraints. The detailed execution of the task may vary with each performance as a result of the changes in the task constraints, but the achievement of the task is still the main aim of the performer. In tennis, for example, the performer is constrained by the rules, they have to control the racket, and they have to change their tactics in response to the actions of their opponent. When serving, the end goal remains the same but the direction of the serve may be adjusted to try and produce errors from your opponent. All of these factors are task constraints.
- *Environmental constraints*: movement takes place in a variety of contexts or environments. Our interaction with the environment constrains our action. Different contexts provide a variety of visual, auditory, and proprioceptive information, all of which act as environmental constraints. The fielder trying to catch the ball in cricket may have a visual difficulty if the sun is in their eyes creating an environmental constraint.

The interaction of these components governs coordination. The performer has a certain level of skill or expertise which is matched to the task in a particular context, over time or in differing contexts; when performing a variety of tasks, this relationship changes. Therefore the performer identifies a functional task goal and by a process of exploration (searches) selects an appropriate solution to the task. This is matched to the skill level of the individual, and learning occurs when a successful achievement of the task takes place.

Part of this search for the most appropriate motor strategy is finding the most appropriate perceptual cues. Hence, both the perception and actions systems are incorporated or mapped into an optimal task solution. Essential to the search for optimal strategies is the exploration of the perceptual-motor workspace. The perceptual-motor workspace is the hypothetical totality of the numerous ways in which perceptual cues and actions are coupled to produce movement. Thus, exploring the perceptual-motor workspace requires searching all possible cues to identify those that are most relevant to the task at hand. Thus, when learning any motor task, there is an exploration stage searching for many solutions to the problem, this is characterized by high variability. As patterns become more stable through practice, there is progressive stability; however, at times there is further instability as this affords a flexible adaptable system (Davids *et al.*, 1997).

Ecological perspective

From an ecological perspective, the work of Gibson (1979) is paramount to the understanding of what information support is necessary for the coordination of actions. Gibson (1979) agued that a human performer becomes specifically attuned to the information available in perceptual flow fields, thus skilful movement is characterized by subtle attunement to the changing demands of the environment. These attunements to changing perceptual invariants become tightly coupled to the intrinsic dynamics of a complex, biological system to act as control parameters. Control parameters can change the macro-organization of the biological system if and when these changes reach critical levels. These changes specify changes afforded to the state of the biomechanical system. Thus, the highly complex interaction between the performer and their environment can be modified and regulated on the basis of small amounts of information

(Davids *et al.*, 1994). The specific movements of the performer create perceptual information to be used to 'guide' the evolving dynamics of the neuromuscular system to the most adaptable pattern required for successful task execution. Thus advocates of the dynamical systems approach to motor control and learning state that an inseparable coupling of perception and action is vital in the generation and improvement of new skills (*Fig. 4*).

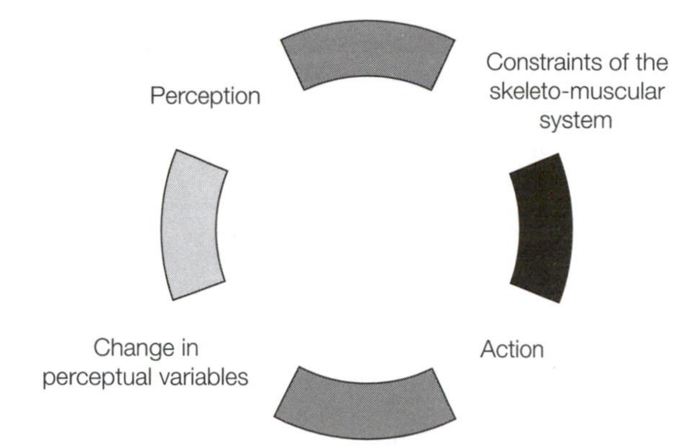

Fig. 4. The coupling of perception and action. Adapted from Williams et al. *(1999).*

Conclusion

Theories of motor learning continue to interest researchers and practitioners, with ongoing debates about the strengths and weaknesses of each theory. An understanding of the theories of learning is a starting point for considering the stages that a learner must pass through to develop skills and also the mode of practice which is most appropriate to practice these skills. In the next section we will address another important concept, the stages of learning.

Further reading

Shea, C.H. and Wulf, G. (2005) Schema theory: a critical appraisal and re-evaluation. *J Mot Behav* **37**, 85–101.

Ulrich, B.D. and Reeve, T.G. (2005) Studies in motor behavior: 75 years of research in motor development, learning, and control. *Res Q Exerc Sport* **76**, S62–70.

Vereijken, B., Van Emmerik, R.E.A., Bongaardt, R., Beek, W.J. and Newell, K.M. (1997) Changing coordinative structures in complex skill acquisition. *Hum Mov Sci* **16**, 823–844.

References

Adams, J.A. (1971) A closed-loop theory of motor learning. *J Mot Behav* **3**, 111–149.

Adams, J.A., Goetz, E.T. and Marshall, P.H. (1972) Response feedback and motor learning. *J Exp Psychol* **92**, 391–397.

Bartlett, F.C. (1932) *Remembering: a study in experimental and social psychology.* Cambridge University Press, Cambridge.

Bernstein, N. (1967) *The Coordination and Regulation of Movements.* Pergamon, Oxford.

Bortoli, L., Robazza, C., Durigon, V. and Carra, C. (1992) Effects of contextual interference on learning technical sports skills. *Percept Mot Skills* **75**, 555–562.

Davids, K., Bennett, S.J., Court, M., Tayler, M.A. and Button, C. (1997) The cognition-dynamics interface. In: Lidor, R. and Bar-Eli, M. (eds) *Innovations in Sport Psychology: linking theory to practice.* ISSP, Netanya, Israel, pp.134–178.

Davids, K., Hanford, C. and Williams, A.M. (1994) The natural physical alternative to cognitive theories of motor behaviour: an invitation for interdisciplinary research in sports science? *J Sports Sci* **12**, 495–528.

Del Rey, P., Wughalter, E.H. and Whitehurst, M. (1982) The effects of contextual interference on females with varied experience in open sport skills. *Res Q Exerc Sport* **53**,108–115.

Dictionary of the Sport and Exercise Sciences. (1991) Human Kinetics, Champaign, IL.

Ericcson, K.A., Krampe, R.T. and Tesch-Romer, C. (1993) The role of deliberate practice in the acquisition of expert performance. *Psychol Rev* **100**, 363–406.

French, K., Rink, E., Judith E. and Werner, P.H. (1990) Effects of contextual interference on retention of three volleyball skills. *Percept Mot Skills* **71**, 179–186.

Gibson, J.J. (1979) *The Ecological Approach to Visual Perception*. Houghton-Mifflin, Boston.

Handford, C., Davids, K., Bennett, S. and Button, C. (1997) Skill acquisition in sport: some applications of an evolving practice ecology. *J Sports Sci* **15**, 621–640.

Head, H. (1926) *Aphasia and Kindred Disorders of Speech*. Cambridge University Press, Cambridge.

Heuer, H. (1991) Invariant relative timing in motor-program theory. In: Fagard, J and Wolff, P.H. (eds) *The Development of Timing Control and Temporal Organisation in Coordinated Action*. Elsevier, Amsterdam, pp.37–68.

Heuer, H., Schmidt, R.A. and Ghodesian, D. (1995) Generalised motor programs for rapid bimanual tasks: a two-level multiplicative-rate model. *Biol Cybern* **73**, 343–356.

Ivry, R.B. (1996) The representation of temporal information in perception and motor control. *Curr Opin Neurobiol* **6**, 851–857.

Kelso, J.A.S. and Norman, P.E. (1978) Motor schema formation in children. *Dev Psychol* **3**, 529–543.

Lai, Q. and Shea, C.H. (1999) Bandwidth knowledge of results enhanced generalized motor program learning. *Res Q Exerc Sport* **70**, 79–84.

Lee, T.D., Magill, R.A. and Weeks, D.J. (1985) Influence of practice schedule on testing schema theory predictions in adults. *J Mot Behav* **17**, 283–299.

Magill, R.A. (2007) *Motor Learning: concepts and applications*, 8th Edn. McGraw-Hill International, New York.

Martenuik, R.G., Leavitt, J.L., MacKenzie, C.L. and Athenes, S. (1990) Functional relationships between grasp and transport components during prehension. *Hum Mov Sci* **9**, 149–176.

McDonald, P., Emmerik, R., and Newell, K. (1989) The effects of practice on limb kinematics in a throwing task. *J Mot Behav* **21**, 245–264.

McCracken, H.D. and Stelmach, G.E. (1977) A test of the schema theory of discrete motor learning. *J Mot Behav* **9**, 193–201.

Newell, K.M. (1974) Knowledge of results and motor learning. *J Mot Behav* **6**, 235–244.

Newell, K.M. (1986) Constraints on the development of coordination. In: Wade, M.G. and Whiting, H.T.A. (eds) *Motor Development in Children: aspects of coordination and control*. Martinus Nijhoff, Boston, pp.341–360.

Newell, K.M. (1991) Motor skill acquisition. *Annu Rev Psychol* **42**, 213–237.

Newell, K.M. (2003) Schema theory (1975): retrospectives and prospectives. *Res Q Exerc Sport* **74**, 383–388.

Newell, K.M., Broderick, M.P., Deutsch, K.M. and Slifkin, A.B. (2003) Task goals and change in dynamical degrees of freedom with motor learning. *J Exp Psychol Hum Percept Perform* **29**, 379–387.

Newell, K.M. and Chew, R.A. (1974) Recall and recognition in motor learning. *J Mot Behav* **6**, 245–253.

Newell, K. and Vaillancourt, D. (2001) Dimensional change in motor learning. *Hum Mov Sci* **20**, 695–715.

Rosenbaum, D.A. (1991) *Human Motor Control*. Academic Press, Inc, San Diego.

Schmidt, R.A. (1975) A schema theory of discrete motor skill learning. *Psychol Rev* **82**, 225–260.

Schmidt, R.A. (2003) Schema theory after 27 years: reflections and implications for a new theory. *Res Q Exerc Sport* **74**, 366–379.

Schmidt, R.A., and Young, D.E. (1987) Transfer of movement control in motor skill learning. In: Cormier, S.M. and Hagman, J.D. (eds) *Transfer of Learning*. Academic Press, Orlando, FL, pp.47–79.

Schmidt, R.A., Zelaznik, H.N., Hawkins, B., Frank, J.S. and Quinn, J.T. Jr. (1979) Motor output variability: a theory for the accuracy of rapid motor acts. *Psychol Rev* **86**, 415–451.

Shapiro, D.C. and Schmidt, R.A. (1982) The schema theory: recent evidence and developmental implications. In Kelso, J.A.S. and Clark, J. (eds) *The Development of Movement Control and Co-ordination*. Wiley, Chichester, pp. 5–78.

Sherwood, D.E. and Lee, T.D. (2003) Schema theory: critical review and implication for the role of cognition in a new theory of motor learning. *Res Q Exerc Sport* **74**, 376–382.

Thelen, E. (1995). Motor development: a new synthesis. *Am Psychol* **50**, 79–95.

Thelen, E., Kelso, J.A.S. and Fogel, A. (1987) Self-organising systems and infant motor development. *Dev Rev* **7**, 39–65.

Turvey, M.T., Fitch, H.L. and Tuller, B. (1982) The Bernstein perspective: 1. The problems of degrees of freedom and context-conditioned variability. In: Kelso, J.A.S. (ed) *Human Motor Behaviour: an introduction*. Erlbaum, Hillsdale, NJ, pp. 239–252.

Van Rossum, J.H.A. (1990) Schmidt's schema theory: the empirical base of the variability of practice hypothesis. *Hum Mov Sci* **15**, 387–435.

Weeks, D.L. and Kordus, R.N. (1998) Relative frequency of knowledge of performance and motor skill learning. *Res Q Exerc Sport* **69**, 224–230.

Williams, A.M., Davids, K. and Williams, J.G. (1999) *Visual Perception and Action in Sport*. Taylor and Francis, London.

Winstein, C.J. and Schmidt, R.A. (1990) Reduced frequency of knowledge of results enhances motor skill learning. *J Exp Psychol Learn Mem Cogn* **16**, 677–691.

Wrisberg, C.A. and Lui, Z. (1991) The effect of contextual variety on the practice, retention and transfer of an applied motor skill. *Res Q Exerc Sport* **62**, 406–412.

Wulf, G., Lee, T.D. and Schmidt, R.A. (1994) Reducing knowledge of results about relative versus absolute timing: differential effects on learning. *J Mot Behav* **26**, 362–369.

Wulf, G. and Schmidt, R.A. (1989) The learning of generalized motor programs: reducing the relative frequency of knowledge of results enhances memory. *J Exp Psychol Learn Mem Cogn* **15**, 748–757.

Wulf, G. and Schmidt, R.A. (1996) Average KR degrades parameter learning. *J Mot Behav* **28**, 371–381.

Wulf, G., Schmidt, R.A. and Deubel, H. (1993) Reduced feedback frequency enhances generalized motor program learning but not parameterization learning. *J Exp Psychol Learn Mem Cogn* **19**, 1134–1150.

Section I

STAGES OF MOTOR LEARNING

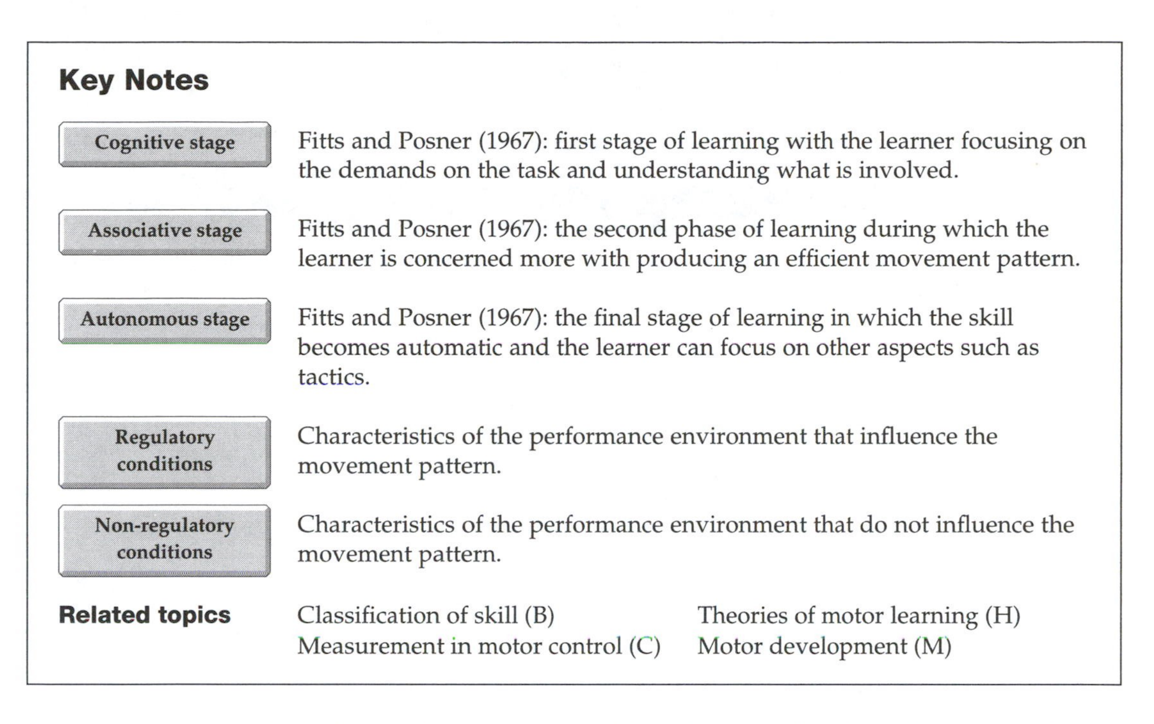

Key Notes

Cognitive stage	Fitts and Posner (1967): first stage of learning with the learner focusing on the demands on the task and understanding what is involved.
Associative stage	Fitts and Posner (1967): the second phase of learning during which the learner is concerned more with producing an efficient movement pattern.
Autonomous stage	Fitts and Posner (1967): the final stage of learning in which the skill becomes automatic and the learner can focus on other aspects such as tactics.
Regulatory conditions	Characteristics of the performance environment that influence the movement pattern.
Non-regulatory conditions	Characteristics of the performance environment that do not influence the movement pattern.

Related topics

Classification of skill (B)
Measurement in motor control (C)
Theories of motor learning (H)
Motor development (M)

In the last section we considered theories of learning. We will now consider the processes of change from an unskilled performer to a skilled performer. When we learn a new skill such as throwing a javelin, or relearn a skill such as riding a bike after an injury, the learning of the skill is a progression. As learning takes place the performer passes through distinct phases. A large body of work is also concerned with the stages a performer goes through when learning or relearning a motor skill. As with theories of control and learning, our understanding of the stages of learning has changed overtime. There has been a move from information-processing perspectives to dynamical/ecological perspectives on the stages of learning. What they all have in common is that they suggest that people go through distinct phases as they learn motor skill. The identification of such stages is an important aspect of coaching or rehabilitation. Several models have been proposed to identify and describe the above stages of motor learning. It is important that the coach, teacher or therapist has a good understanding of the phases of learning that a performer will go through. The next part of this section presents an overview of the more influential work in this area.

> *Box 1. Power law of practice*
>
> In 1926 Snoddy created a mathematical law that can be used when looking at changes in the rate of improvement during skill acquisition. This law states that in early practice there are large improvements, but after this rapid improvement the rate slows. Changes in the rate of learning are dependent upon the task and the experience of the performer.

The Fitts and Posner three-stage model

In 1967 Fitts and Posner presented a model to account for the frequently observed stages of learning, using the concept of hierarchical organization (this model was in fact first proposed by Fitts in 1964). The model of Fitts and Posner consists of three stages – the cognitive, associative or intermediate and the autonomous stage of motor learning. As learning progresses one phase gradually merges into another, so that there is little definite transition between them (*Fig. 1*).

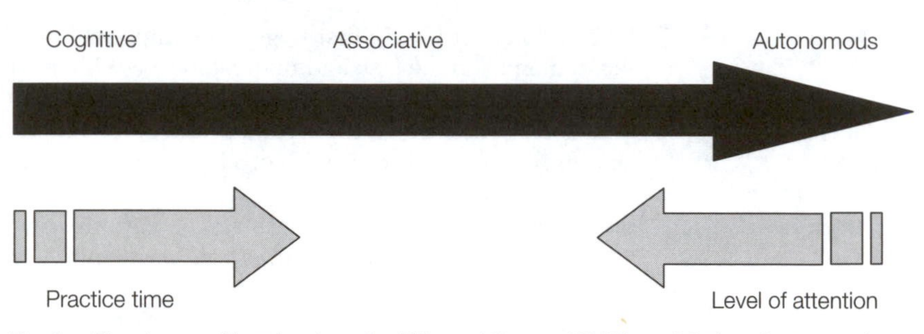

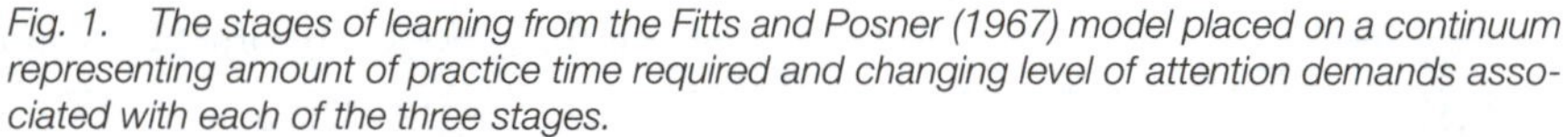

Fig. 1. The stages of learning from the Fitts and Posner (1967) model placed on a continuum representing amount of practice time required and changing level of attention demands associated with each of the three stages.

Cognitive stage

Imagine you are watching a novice hit a cricket ball. When first learning to intercept the ball the novice has to concentrate very hard, attending to many if not all aspects of the task. The novice would initially make lots of errors and miss the ball frequently as they try to experiment with different movement strategies. What you are observing is someone in the first stage of learning: the cognitive stage. In this stage, the learner is concerned with understanding the task, asking themselves lots of questions and attending to many cues, events and responses that later go unnoticed. At this early stage of learning, instructions and demonstrations are most effective, allowing the selection of the initial repertoire of subroutines. In therapy, the use of facilitation techniques and/or manual guidance is required to show the patient what needs to be done. In addition, some learners engage in self-talk, verbally guiding themselves through actions (Schmidt, 1991). Performance at this stage is marked by a large number of errors with high variability and little consistency. However, improvements in performance are also quite large, perhaps as a result of selecting the most effective strategy for the task.

Associative/intermediate stage

When entering the second stage, the associative stage of learning, the cricketer would now be able to make contact with the ball and would begin to extract cues from the environment, such as the speed of the bowler, to refine their movement strategy. In addition, the ball can now be played to different areas of the pitch as the relationship between the position of their feet prior to contact and the direction of the shot is understood. Similarly, in therapy the patient is capable of detecting errors or solving problems without the therapist. This second stage of learning lasts for varying periods, dependent upon the complexity of what is to be learnt (Fitts and Posner, 1967). Several factors change markedly during the associative stage. The performer has to learn to associate certain environmental cues with the movements that are required to achieve the goal of the skill (perception–action coupling). In particular, performers discover cues for effective timing of the action and anticipatory skills dramatically improve, making movements smoother and better timed. Associated with more efficient movement patterns are improvements in performance; fewer errors are made, but some inter-trial variability is still apparent as the performer explores new solutions to the movement problems posed. Enhanced movement efficiency also reduces energy costs, and self-talk become less important for performance.

Autonomous stage

At the autonomous stage the skill is almost automatic, or habitual. The component processes become increasingly autonomous, less directly subject to cognitive control and to interference from other ongoing activities and environmental distractions (Fitts and Posner, 1967). The cricketer in this phase of learning now requires less processing to produce a skilled on-drive and it can now be carried out while new learning is in progress or while the individual is involved in other cognitive or perceptual activity. For example, while executing the shot the cricketer can now search the pitch to identify fielders, which will subsequently affect how and where the final shot is driven. In this stage of learning, the speed and efficiency with which some skills are performed continue to increase. Performance variability is low during this stage; the skill is performed consistently from one attempt to the other. Additionally, at this stage most performers can detect their own errors, and make proper adjustments to correct them. The likelihood of a performer reaching the final stage of motor learning is highly dependent upon the quality of instruction and practice.

Fitts and Posner (1967) make use of components of information-processing theories, such as sensory input and attention, and value the implementation of mental and observational practice and using feedback to enhance learning. Although it is very difficult to ascertain when a performer enters each of the three stages, Fitts and Posner's (1967) description of motor learning illustrates very well how motor behavior can change as a function of practice and/or instruction. Consequently, this provides the therapist and/or instructor with valuable clues that can help them develop appropriate practice schedules in appropriate settings. Langley (1995) looked at learning to bowl by making observations and interviewing participants at the beginning, middle and end of a 10-week programme. It was found that during the early stages of learning participants focused on the fact that they felt they had lack of control of the ball. In the middle of the 10 weeks they started to describe the movement characteristics, and by the end they were concerned with consistency and accuracy. This shows a progression in skill development that can be compared to models of stages of learning such as those of Fitts and Posner (1967) and Gentile (1972).

Ecological theorists would argue that while the cognitive approach seeks to identify the 'laws of learning' and use these to explain the stages of learning a motor skill, it is more beneficial to identify how the organism 'learns the laws' intrinsic to the task environment (Vereijken, 1991). A common framework used by ecological theorists to explain skill acquisition and motor coordination is Bernstein's stages of learning (1967). Similar to Fitts and Posner this framework focuses on three stages in which the performer learns a motor skill. This theory suggests that at first a basic coordination pattern is established as degrees of freedom are reduced to a minimum and through practice, thus allowing the performer to try and solve the movement problem in many ways; an efficient, modifiable and flexible movement solution emerges.

Bernstein's stage theory of motor learning

According to Bernstein (1967), the learning of a motor skill can be likened to the solving of a problem, the problem being how best to harness the many available degrees of freedom of the human movement system. Degrees of freedom are all available muscles, joints, tendons and ligaments of the human body which are free to vary and thus must be controlled for purposeful movement and give rise to Bernstein's (1967) degrees of freedom problem.

Bernstein (1967) highlighted that the complex interactions between the different components of the movement apparatus (e.g. muscles, tendons and joints) make separate control of all these components impossible. Therefore, the large number of individual degrees of freedom has to be reduced to make control possible. Bernstein (1967) claimed that the main concern for a beginner in a motor task is to master the multiple and redundant degrees of freedom potentially involved; he contended that coordination itself was 'the process of mastering redundant degrees of freedom of the moving organ, in other words its conversion to a controllable system' (Bernstein, 1967, p.127). He proposed that learning took place in three stages:

1. *novice stage*: freezing out a portion of the degrees of freedom
2. *advanced stage*: reinstating degrees of freedom
3. *expert stage*: releasing, reorganizing, and exploiting degrees of freedom.

Novice stage: freezing degrees of freedom

Initially in skill learning, the central control mechanisms cannot functionally solve the degrees of freedom problem in the sense of producing a stable, task-relevant coordination solution. Bernstein (1967) speculated that control entails a process of initially 'freezing out' degrees of freedom that are subsequently harnessed or coupled to form a synergy or coordinative structure (Turvey, 1977, 1990). Following Bernstein (1967), it has been suggested that the mastering of the degrees of freedom can be achieved by following two alternative and/or complementary strategies. The first one consists of 'freezing' a number of joints, thus reducing the number of active degrees of freedom, i.e. temporarily freezing degrees of freedom. This rigid and inflexible functioning is also obvious in the early stages of learning many dynamic movements, such as catching with two hands (Utley *et al.*, 2007) or one hand (Davids *et al.*, 1997), kicking a football (Davids *et al.*, 2000) or a volleyball strike (Temperado *et al.*, 1997).

Vereijken *et al.* (1992) conducted an empirical test of the freezing of degrees of freedom in adults learning a ski simulator task (see Research Highlight, p.126). If the degrees of freedom at a joint are 'frozen', that signifies that virtually no movement is occurring, which can be observed by a reduction of angular excursion of a joint; as Vereijken *et al.* (1992) suggested, this is operationally defined in terms of the minimization of the standard deviation of joint angles. Kinematic analysis of lower limb joints and the torso motions clearly showed, consistent with Bernstein's (1967) hypothesis, that during initial learning of a novel task the learner freezes degrees of freedom, in an attempt to maintain balance on the apex of the simulator.

The second strategy, drawing upon the original contribution of Bernstein (1967), consists of introducing rigid couplings between the various articular degrees of freedom, which are constrained to act as one functional unit (*Fig.* 2). These functional units are referred to as coordinative structures (Turvey, 1977); they are tuned to function specifically in each unique task by environmental information. McDonald *et al.* (1989) compared intra-limb coupling of the

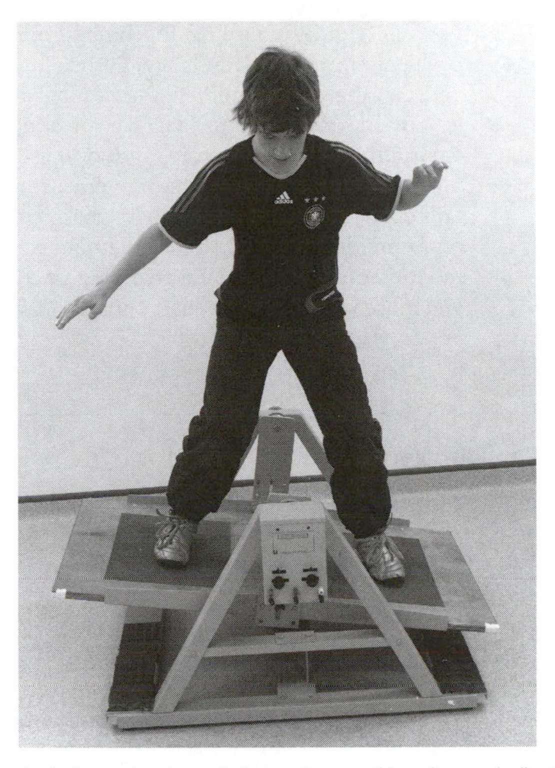

Fig. 2. Learning to balance involves all three stages of learning as indicated by Bernstein (1967).

preferred and non-preferred hand while learning to throw darts. Cross-correlations between joint displacements were examined under the assumption that high cross-correlations are a reflection of a tight coupling between constituent parts. McDonald *et al.* (1989) reported relatively consistent hand trajectories for either hand, while the inter-joint correlations underlying these were significantly lower in the dominant arm, although both limbs demonstrated high coupling. Again these results confirmed that the degrees of freedom of the movement system are not controlled independently but as a single functional unit, and that more degrees of freedom are freed in the dominant arm throughout learning.

In addition to dependencies within-limb, coordinative structures are formed across the limbs. Vereijken *et al.* (1992) examined Bernstein's propositions in novices practicing slalom-like ski movements on a ski simulator (see Research Highlight). Results showed that early in practice trials, the joint angles of the lower limbs displayed very little movement (i.e. a small range of motion), whereas at the same time the interjoint couplings were high. The results purported by Vereijken and colleagues confirm Bernstein's (1967) notion that the degrees of freedom of an unskilled movement system are not controlled independently but as a unit to reduce degrees of freedom and aid coordination. These findings have been extended to other tasks as well, such as reaching a target (Kelso *et al.*, 1979, 1983), handwriting (Newell and van Emmerik, 1989; van Emmerik and Newell, 1990) volleyball serving (Temperado *et al.*, 1997) and catching (Astill and Utley, 2006; Tayler and Davids, 1997).

While rigidly coupling the limbs appears to be an elegant solution to the problem of motor redundancy, Bernstein (1967) noted that a balance has to exist between the need to reduce the number of degrees of freedom to be controlled at any one time and the facility for adaptation in an ever-changing dynamic environment. Subsequently, individuals release degrees of freedom and the inherent redundancy of the movement system leads to a search for the most appropriate coordination mode, many motor solutions are explored and the performer learns to select the adequate strategy for a particular task. Such exploration is often shown by a releasing of the degrees of freedom (Ko *et al.*, 2003) and more variable movement patterns (Touwen, 1998; Davids *et al.*, 2000).

Research Highlight: Vereijken *et al.* (1992)

Subjects practiced on a ski simulator on which they had to move from side to side making slalom-like ski movements. Each subject completed a total of 140 trials of 1 minute duration for 7 consecutive days. Kinematics were recorded from 11 light-emitting diodes (LEDs). Task performance data were collected and changes in the individual degrees of freedom as a function of practice examined.

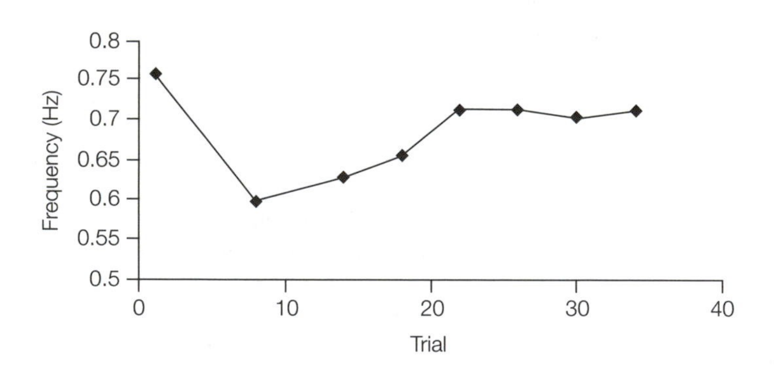

Fig. 3. Mean frequency of movement. Reproduced with permission from Vereijken et al. *(1992).*

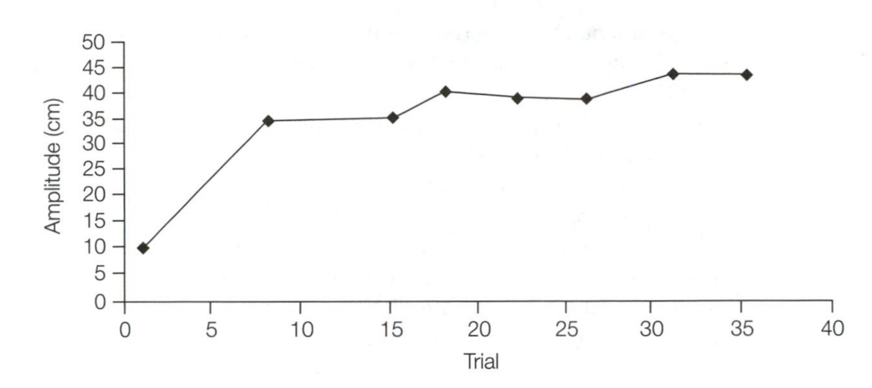

Fig. 4. Mean amplitude of movement. Reproduced with permission from Vereijken et al. *(1992).*

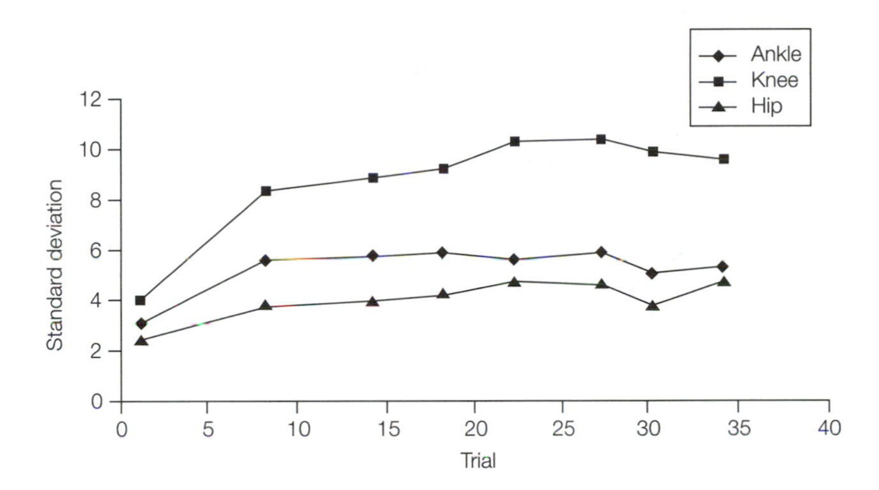

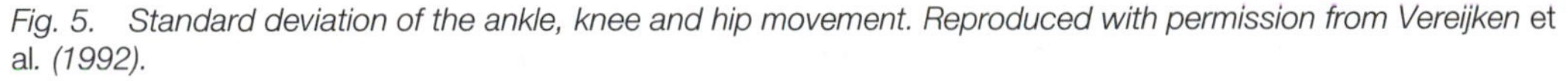
Fig. 5. Standard deviation of the ankle, knee and hip movement. Reproduced with permission from Vereijken et al. *(1992).*

Changes in performance can be seen in *Fig. 3* and *4*, with subjects initially moving at a high frequency with small amplitude, and over trials the subjects increased both frequency and amplitude. *Figure 5* shows a gradual release in the degrees of freedom as the average standard deviation increases.

Advanced stage: reinstating degrees of freedom

Bernstein suggested that freezing degrees of freedom is followed, as a consequence of 'experiment and exercise', by 'lifting of all restrictions, that is, to the incorporation of all possible degrees of freedom' (p.108). In Bernstein's view, practice affords learning to control the potentially complicating reactive phenomena of the forces of one body segment influencing another. In this stage of learning 'a greater economy of movement and diminution of fatigue' is evident, in contrast to the 'high degree of expensive titanic fixation' observed before this. By this second stage of learning the relationship between body parts and the coordination of basic actions with environmental objects and surfaces has been established. The challenge to the performer is now to discover the laws that govern their control. The performer becomes more perceptually tuned to the consequences of different combinations of perceptual cues to channel the search through dynamics of the movement (Fowler and Turvey, 1978). Fitch *et al.* (1982) state that the consequences associated with the exploration of the workspace act in tune with action systems until the desired kinematic outcomes are attained and task constraints satisfied.

Exploration of the workspace is only made possible if the couplings assembled early in learning are progressively released and the relationship between the limbs becomes more independent. Thus part of the ban on the degrees of freedom is gradually released and the formerly 'frozen' degrees of freedom become more incorporated into the movement system. Thus learning is characterized by an increase in the number of active degrees of freedom and an increase in the amplitude of articular movements. The speed and nature of the release of the ban on degrees of freedom in adult motor learning varies with the task being learned and, more importantly, with the constraints on action (Newell and Vaillancourt, 2001). An early demonstration of the releasing pattern was demonstrated by Arutyunyan *et al.* (1968) and later by Southard and Higgins (1987). Southard and Higgins showed that there was a progressive release of the distal arm joint segments in the learning of a type of racquetball forehand shot. Other studies found only weak or no evidence of releasing of degrees of freedom (McDonald *et al.*, 1989; Newell and van Emmerik, 1989, respectively). In a more recent study Broderick and Newell (1999) noted a different pattern of change to that noted previously in subjects learning to bounce a basketball. It was found that the progression of skill in basketball bouncing was paralleled by motions of the arm segments that in the beginner were both proximal (shoulder) and distal (wrist) and that, with further practice, moved toward active control of the effector chain, the elbow. It thus seems that open-chain linkages are more prone to exhibit a proximal-to-distal direction to the bringing in of the biomechanical degrees of freedom, but this pathway of change is due to the particular task constraints and is not a general learning strategy.

Results from Vereijken *et al.* (1992) offer support for release on the ban on the degrees of freedom, as after seven days of practice on the ski simulator the 'rigidly fixed joint angles' became 'more active participants in the performance' and an increase in the variability of the joint angles was noted. If we accept that learning is a process of continually searching for regions of stability or attraction within each individual workspace, it can be deduced that variability in movement dynamics enhances the search process. As Vereijken *et al.* (1992) have shown, in tasks involving multiple degrees of freedom within a dynamical environment context, an initial rigid coupling between system components gives way to a more flexible regime characterized by an increase in variability in joint motion. Thus variability may actually be indicative of a functional level of independence between components of the body that all contribute towards the achievement of a task goal.

As degrees of freedom are released there is often a differentiation in the patterns of coordination between the two body sides. By freezing degrees of freedom in one limb, the other is limb free to explore its workspace (Newell *et al.*, 1989). The work of McDonald *et al.* (1989) showed that the inter-joint correlations of the non-dominant limb were significantly lower than those in the dominant arm, although both limbs demonstrated high coupling. In contrast, Newell and van Emmerik (1989) showed the opposite: the non-dominant limb showed lower inter-joint correlations. These results suggest a progressive releasing of the degree of freedom in either the dominant or non-dominant arm to search for new coordination codes, suggesting that handedness is task specific (Peters, 1977). Similarly, asymmetries between body sides of the lower limbs were noted by Vereijken *et al.* (1992). The largest decreases in cross-correlations seen in the left limb were consistent with an increase in cross-correlations in the right limb. It seems that raising the ban on degrees of freedom in one limb is easier for performers to deal with.

Expert stage: releasing, reorganizing and exploiting the degrees of freedom

The final stage of Bernstein's learning framework corresponds to 'a degree of coordination at which the organism is not only unafraid of reactive phenomena in a system with many degrees of freedom, but is able to structure its movements so as to "utilise entirely the reactive phenomena which arise"' (Bernstein, 1967, p.109). Thus to attain skill, components of the control structure should be quantitatively scaled so that the reactive forces of the limbs become largely responsible for movement. Skill is exhibited when exploitation of passive, inertial and mechanical properties of limb movements occurs. Additional degrees of freedom may be released as coordinative structures become more stable and there seems to be fluidity of movement (Bernstein, 1967). Researchers specifically addressed this third stage of motor learning by studying slalom-like movements on a ski simulator (Vereijken *et al.*, 1992) and energy expenditure during learning the ski simulator task (Durand *et al.*, 1994). Each of these studies indicated that a novice becomes more efficient at generating muscular force by exploiting the available passive (reactive, frictional and inertial) forces.

The first study supporting Bernstein's third stage of motor learning examined rapid arm movements (Schneider *et al.*, 1989). It was found that following practice, subjects were able to use the muscle moments (active forces) as complementary forces to the passive interactive components of the moving limb. This exploitation of the reactive forces was evident in all segments of the limb motion, and was particularly apparent at the point of movement reversals. The authors concluded that these results provide direct evidence of the utilization of reactive phenomena as an advanced feature of skilled performance. However, more recently Spencer and Thelen (1999) have shown that enhanced co-contraction is a learned phenomenon in the same task, a feature that seems counter to the general notion of a release of degrees of freedom and the reduction of limb stiffness with learning.

The hypothesis that the degrees of freedom that are frozen during initial learning are then released requires further empirical testing (Newell and Vaillancourt, 2001). In spite of demonstrations of coordination change with practice supporting Bernstein's (1967) contentions, there is growing evidence that there may not be, as suggested by Bernstein, a single pathway of change or a universal learning strategy. Conversely, it is suggested that the combined notions of initially freezing and releasing degrees of freedom are constraint- and in particular task-dependent, and not general directional stage-like strategies of learning to harness the degrees of freedom (Newell, 1986; Newell and McDonald, 1994; Newell and Vaillancourt, 2001; Ko *et al.*, 2003). Thus we have, so far, little understanding of the search strategies that are used to change the number of degrees of freedom that are independently regulated in the coordinated solution, or of the specific role of task constraints in the process (Newell *et al.*, 1989).

Research Highlight: Brain research

Over the last 10 years there has been an increase in the use of techniques such as functional magnetic resonance imaging (fMRI) and positron emission tomography (PET) scans by researchers interested in changes in brain activity during the learning of motor skills. It would appear that areas of the brain that are active during the early stages of learning a skill are not always active in the later stages of skill acquisition. Studies of learning have demonstrated frontoparietal, subcortical, and cerebellar involvement. However, there is still conflicting evidence on the contribution made by individual regions of the brain during learning. One problem is caused by limitations in technology. Scanning devices cannot currently be used on dynamic or serial tasks. Many of the studies conducted so far are based on discrete finger or hand movements often viewed on a computer. The work of Doyon and Ungerleider (2002) has proposed that two distinct circuits are most commonly associated with skill acquisition. These areas are the cortico-basal ganglia-thalamo-cortical loop and the cortico-cerebello-thalamo-cortical loop. It would appear that early in learning the cortico-cerebello-thalamo-cortical loop is more involved. Advances in technology will bring more studies of this nature and will add to our understanding of motor learning. Other studies of interest in this area include those of Muller *et al*. (2002) and Parsons *et al*. (2005).

Gentile's two-stage model

The most recent model concerned with stages of motor learning and development is that proposed by Gentile (1972, 1987, 2000). In contrast to Fitts and Posner, Gentile viewed motor skill learning as passing through two stages; these stages are presented from the perspective of the goal of the learner. Gentile considers how the nature of the movement environment in which the task has to be performed influences the nature of information a learner must acquire. She also highlights a number of instructional strategies that a teacher can implement to facilitate a learner's progress through the two stages.

Getting the idea of the movement

In the first stage, the goal of the learner is 'getting the idea of the movement' (Gentile, 1998). This first stage of learning refers to the need for the learner to establish an appropriate coordination pattern. Thus, the performer puts into place a movement structure that they consider appropriate and is good enough to meet task demands. With respect to learning to catch, the focus in the first stage of learning is acquiring the appropriate arm and hand coordination that will lead to successful interception, if not a catch. This involves an explicit process whereby there is a conscious mapping of a set of correspondence between the performer and the environment conditions in order to achieve the task goal.

In addition to establishing the basic movement pattern, the performer must also learn to discriminate between environmental features that specify how the movements must be produced and those that do not influence performance. Gentile (2000) referred to these features as regulatory or non-regulatory environmental conditions. Regulatory conditions are characteristics of the performance environment that influence the movement pattern. For example, when catching a ball, such characteristics include the size of the ball, the distance the ball is from the performer and the trajectory of the ball. With respect to rehabilitation procedures Gentile (2000) cautions that, when handling techniques or manual guidance are used, the therapist becomes part of the regulatory conditions; this alters intrinsic feedback available to the patient. This in turn may interfere with the initial coordination pattern assembled.

Similar to Fitts and Posner's cognitive stage of learning, Gentile states that in order to achieve the action goal, a large amount of cognitive problem solving is undertaken. In order for the initial movement coordination pattern to be assembled and the performer to distinguish between regulatory and non-regulatory conditions, the learner has to explore a variety of movement possibilities through trial and error. Once the initial coordination pattern is produced that allows the performer to solve the task, it is unlikely that it is efficient or consistent; however, as Gentile (2000) states, the learner 'does have a framework for organizing an effective movement' (p.119).

Fixation and diversification

In the second stage of learning, the learner's goal is described as fixation/diversification. During this stage the learner must acquire several characteristics to continue skill improvement. Firstly, the learner must develop the capability of adapting the movement pattern, so it is commensurate with specific demands of any performance situation. Secondly, the person must increase his/her consistency in achieving the goal of the skill. Lastly, the learner must be able to produce the desired coordination pattern with an economy of effort. These changes in the organization of coordination pattern are not conscious but are considered to result from learning (Gentile, 1998).

A unique feature of Gentile's model is that the learner's goals depend upon the type of skill, more specifically if it is an open or closed skill. As discussed in Section B, a closed skill is a motor skill that is performed in a stable or predictable environment where the performer determines when to begin the action. Closed skills, according to Gentile, require fixation of the basic movement-coordination pattern. This means that the learner must refine the pattern so that the action goal can be achieved correctly, consistently and efficiently from trial to trial. This is not to say that the performer does not need to make modifications to the pattern; however, the invariant features of the movement pattern itself will not change but the parameters of the movement may. In contrast, open skills require diversification of the basic movement pattern acquired during the first stage of learning. An important characteristic of open skills that differs from closed skills is the requirement for the performer to quickly adapt to the continuously changing spatial and temporal regulatory conditions of the skill. This means that the learner must become attuned to critical features of the regulatory conditions and acquire the capability to modify the movement patterns to meet the constantly changing demands placed upon the performer.

The three descriptions of learning outlined in this section enhance our knowledge of the learning process in a number of different ways. Fitts and Posner's model helps us understand the cognitive activity and associated behavioral processes that characterize the learner through their three stages of learning. Bernstein's model attempts to explain how the novice solves the 'degrees of freedom problem' and illustrates the importance of the changing relationship between the performer and their dynamics of action, their perceptual activity and their environment and how all contribute to the learning process. Similarly, Gentile considers the nature of the environment in which a skill is ultimately performed, how this influences the goal of the performer, and how the coach and/or therapist can help the performer meet those goals. It is the acknowledgment of the importance of the environment and how this influences learning which enables a coach and/or therapist to really understand how a skill can be learnt. Thus, in addition to providing an insight into the learning process, studying stages of learning also provides us with a wealth of information that can be used to structure practice sessions to help foster the many movement skills required in the diverse environmental contexts they are performed in. The implications for coaching, rehabilitation, and development will be considered in depth in Section K.

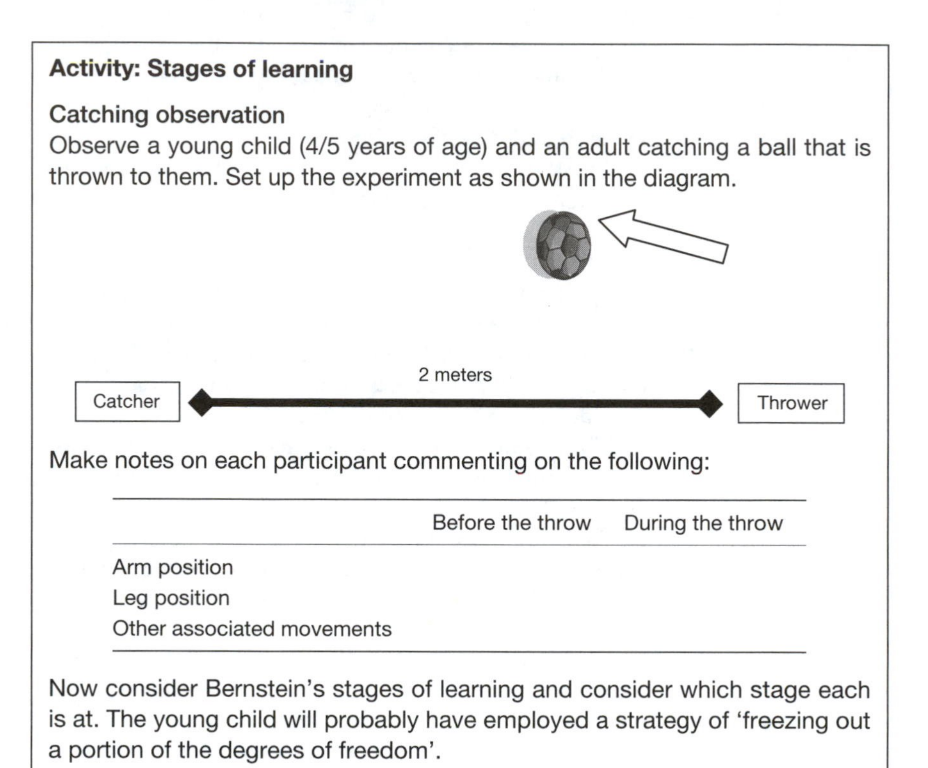

Activity: Stages of learning

Catching observation

Observe a young child (4/5 years of age) and an adult catching a ball that is thrown to them. Set up the experiment as shown in the diagram.

Make notes on each participant commenting on the following:

	Before the throw	During the throw
Arm position		
Leg position		
Other associated movements		

Now consider Bernstein's stages of learning and consider which stage each is at. The young child will probably have employed a strategy of 'freezing out a portion of the degrees of freedom'.

Indicators of learning

Researchers have conducted numerous studies in order to examine the stages of learning during skill acquisition and the changes in coordination patterns. When we are learning a new skill we are learning to control and coordinate multiple degrees of freedom. A variety of methods have been employed in order to examine learning when performing a range of tasks that vary in complexity. Researchers have collected a variety of data; including kinematic data, EMG data, force data, and fMRI data in order to better understand learning. Ko *et al.* (2003) looked at how participants learnt to coordinate redundant degrees of freedom in a dynamic balance task. They found that there were changes over time and that these changes were influenced by the task. In the initial phases of learning the motion of the torso and limb segments were less coherent as the performers attempted to balance. Unlike the participants in the study of Vereijken *et al.* (1992), the nature of the task involved moving towards a balance position which by its very nature would elicit few large movements but many small adjustments. This gives the impression that the performer is freezing the degrees of freedom rather than releasing them as proposed by Bernstein. Ko *et al.* (2003) state that this is indicative of the performers searching for a more 'stable and efficient postural control mode' (p.63). Therefore, due to the nature of the task the participants release degrees of freedom during the early stages of learning and as they practice use pure ankle control or hip control depending on the motion of the board.

A number of studies have provided evidence of Bernstein's (1967) three stages of learning in a variety of tasks. These include:

- McDonald *et al.* (1989), dart throwing
- Southard and Higgins (1987), racquetball
- Broderick and Newell (1991), ball bouncing
- Beek and van Santvoord (1992), juggling
- Vereijken *et al.* (1992), skiing
- Chen *et al.* (2005), pedalo locomotion
- Hong and Newell (2006), skiing
- Utley *et al.* (2007), catching.

Another interesting study was conducted by Roux *et al.* (2006). They examined shoulder movements in participants who were learning to use a wheelchair. One of the main aims of this study was to test Bernstein's stages of learning theory by kinematic analysis of the motion of the shoulder in novice wheelchair users. They expected to find freezing of the degrees of freedom during the early phases of learning, with a gradual release as learning occurred. However, they did not observe any 'freezing' of the degrees of freedom but they did observe 'unfreezing' of the degrees of freedom as the propulsion pace and amplitude of joint rotation increased over trials. The authors conclude that the complexity of the task was relatively low and therefore participants found the task simple in the early stages of learning. As with all studies that have examined learning, there is great variation in findings; this is often due to the nature of the task and the context in which it is performed having a significant influence on the results.

Conclusion

When we learn new motor skills we pass through a number of stages on a continuum from novice to expert. There are a number of models that have been presented to explain the stages that a learner goes through, and over the years researchers have attempted to test these models in a variety of contexts in both the sporting and rehabilitation context. It is vital that anyone engaged in teaching, coaching, or rehabilitation gains an understanding of these stages and considers the implications for skill acquisitions. We will look in depth at the learning context, modes of practice and retention in Section K.

Further reading

Haibach, P.S., Daniels, G.L. and Newell, K.M. (2004) Coordination changes in the early stages of learning to cascade juggle. *Hum Mov Sci* **23**, 185–206.

Luft, A.R. and Buitrago, M.M. (2005) Stages of motor skill learning. *Mol Neurobiol* **32**, 205–216.

Müller, R.A., Kleinhans, N., Pierce, K., Kemmotsu, N. and Courchesne, E. (2002) Functional MRI of motor sequence acquisition: effects of learning stage and performance. *Brain Res Cogn Brain Res* **14**, 277–293.

Shea, C.H., Park, J.H. and Braden, H.W. (2006) Age-related effects in sequential motor learning. *Phys Ther* **86**, 478–488.

Vereijken, B., van Emmerik, R.E.A., Bongaardt, R., Beek, W.J. and Newell, K.M. (1997) Changing coordinative structures in complex skill acquisition. *Hum Mov Sci* **16**, 823–844.

References

Arutyunyan, G.H., Gurfinkel, V.S. and Mirskii, M.L. (1968) Investigation of aiming at a target. *Biophysics* **13**, 536–538.

Astill, S.L. and Utley, A. (2006). Two-handed catching in children with developmental coordination disorder. *Motor Control* **10**, 109–124.

Beek, P.J. and van Santvoord, A.A.M. (1992) Learning the cascade juggle: a dynamical systems analysis. *J Mot Behav* **24**, 85–94.

Bernstein, N. (1967) *The Coordination and Regulation of Movements.* Pergamon, Oxford.

Broderick, M.P. and Newell, K.M. (1999) Coordination patterns in ball bouncing as a function of skill. *J Mot Behav* **31**, 165–189.

Chen, H.H., Liu, Y.T., Mayer-Kress, G. and Newell, K.M. (2005) Learning the pedalo locomotion task. *J Mot Behav* **37**, 247–256.

Davids, K., Bennett, S.J., Court, M., Tayler, M.A. and Button, C. (1997) The cognition–dynamics interface. In: Lidor, R. and Bar-Eli, M. (eds) *Innovations in Sport Psychology: linking theory to practice*. ISSP, Netanya, Israel, pp.762–789.

Davids, K., Lees, A. and Burwitz, L. (2000) Understanding and measuring coordination and control in soccer skill: implications for talent identification and skill acquisition. *J Sports Sci* **18**, 703–714.

Doyon, J. and Ungerleider, L. (2002) Functional anatomy of motor skill learning. In: Squire, L. and Schacter, D. (eds) *Neuropsychology of Memory*. Oxford University Press, Oxford, pp.225–238.

Durand, M., Geoffroi, V., Varray, A. and Prefaut, C. (1994) Study of the energy correlates in the learning of a complex self. *Hum Mov Sci* **13**, 785–799.

Fitch, H.L., Tuller, B. and Turvey, M.T. (1982) The Bernstein perspective: III. Tuning coordinative structures with special reference to perception. In: Kelso, J.A.S. (ed) *Human Motor Behaviour: an introduction*. Hillsdale, NJ, pp. 271–281.

Fitts, P.M. (1964) The information capacity of the human motor system in controlling the amplitude of movement. *J Exp Psychol* **47**, 381–391.

Fitts, P. and Posner, M.I. (1967) *Human Performance*. Brooks/Cole Publishing, Belmont, CA.

Fowler, C.A. and Turvey, M.T. (1978) Skill acquisition: an event approach for the optimum of a function of several variables. In: Stelmach, G.E. (ed) *Information Processing in Motor Control and Learning*. Academic Press, New York, pp.590–598.

Gentile, A.M. (1972) A working model of skill acquisition with application to teaching. *Quest Monograph* **17**, 3–23.

Gentile, A.M. (1987) Skill acquisition: action, movement, and neuromotor processes. In: Carr, J.H., Shephard, R.B., Gordon, J., Gentile, A.M. and Held, J.M. (eds) *Movement Science: foundations for physical therapy in rehabilitation*. Aspen, Rockville, MD, pp.93–154.

Gentile, A.M. (1998) Implicit and explicit processes during acquisition of functional skills. *Scand J Occup Ther Pediatr* **20**, 29–50.

Gentile, A.M. (2000) Skill acquisition: action, movement, and neuromotor processes. In: Carr, J.H. and Shephard, R.B. (eds) *Movement Science: foundations for physical therapy*, 2nd Edn. Aspen, Rockville, MD, pp.111–187.

Hong, S.L. and Newell, K.M. (2006) Change in the organization of degrees of freedom with learning. *J Mot Behav* **38**, 88–100.

Kelso, J.A.S., Putnam, C.A. and Goodman, D. (1983) On the space-time structure of human interlimb coordination. *Q J Exp Psychol Hum Percept Perform* **35**, 347–375.

Kelso, J.A.S., Southard, D.L. and Goodman, D. (1979) On the coordination of two-handed movements. *J Exp Psychol Hum Percept Perform* **5**, 229–238.

Ko, Y.G., Challis, J.H. and Newell, K.M. (2003) Learning to coordinate redundant degrees of freedom in a dynamic balance task. *Hum Mov Sci* **22**, 47–66.

Langley, D.J. (1995) Student cognition in the instructional setting. *J Teach Phys Educ* **15**, 25–40.

McDonald, P., Emmerik, R. and Newell, K. (1989) The effects of practice on limb kinematics in a throwing task. *J Mot Behav* **21**, 245–264.

Muller, R., Kleinhans, N., Pierce, K., Kemmotsu, N. and Courchesne, E. (2002) Functional MRI of motor sequence acquisition: effects of learning stage and performance. *Brain Res Cogn Brain Res* **14**, 277–293.

Newell, K.M. (1986) Constraints on the development of coordination. In: Wade, M.G. and Whiting, H.T.A. (eds) *Motor Development in Children: aspects of coordination and control.* Martinus Nijhoff, Dordrecht, pp.341–360.

Newell, K.M., Kugler, P.N., van Emmerik, R.E.A. and McDonald, P.V. (1989) Search strategies and the acquisition of coordination. In: Wallace, S.A. (ed) *Perspectives on Coordination*. North Holland, Amsterdam, pp.85–122.

Newell, K. and McDonald, P.V. (1994) Learning to coordinate redundant biomedical degrees of freedom. In: Swinnen, S., Heuer, H., Massion, J. and Casaer, P. (eds) *Interlimb Coordination. Neural, dynamical, and cognitive constraints.* Academic Press, San Diego, pp.515–536.

Newell, K. and Vaillancourt, D. (2001) Dimensional change in motor learning. *Hum Mov Sci* **20**, 695–715.

Newell, K.M. and van Emmerik, R.E.A. (1989) The acquisition of coordination: preliminary analysis of learning to write. *Hum Mov Sci* **8**, 17–32.

Parsons, M.W., Harrington, D.L. and Rao, S.M. (2005) Distinct neural systems underlie learning. *Hum Brain Mapp* **24**, 229–247.

Peters, M. (1977) Simultaneous performance of two motor activities: the factor of timing. *Neuropsychologia* **15**, 461–465.

Roux, L., Hanneton, S. and Roby-Brami, A. (2006) Shoulder movements during the initial phase of learning manual wheelchair propulsion in able-bodied subjects. *J Clin Biomech* **21**, S45–51.

Schmidt, R.A. (1991) *Motor Learning and Performance: from principles to practice.* Human Kinetics, Champaign, IL.

Schneider, K., Zernicke, R.A., Schmidt, R.A. and Hart, T.J. (1989) Changes in limb dynamics during practice of rapid arm movements. *J Biomech* **22**, 805–817.

Snoddy, G.S. (1926) Learning and stability: a psychological analysis of a case of motor learning with clinical applications. *J Appl Psychol* **10**, 1–36.

Southard, D. and Higgins, T. (1987) Changing movement patterns: effects of demonstration and practice. *Res Q Exerc Sport* **58**, 77–80.

Spencer, J.P. and Thelen, E. (1999) A multimuscle state analysis of adult motor learning. *Brain Res Exp Brain Res* **128**, 505–516.

Tayler, M.A. and Davids, K. (1997) Catching with both hands: an evaluation of neural cross-talk and coordinative structure models of bimanual coordination. *J Mot Behav* **23**, 263–279.

Temperado, J.J., Della-Grasta, M., Farrell, M. and Laurent, M. (1997) A novice-expert comparison of (intra-limb) coordination subserving the volleyball serve. *Hum Mov Sci* **16**, 653–676.

Touwen, B.C.L. (1998) The brain function and development of function. *Dev Rev* **18**, 504–526.

Turvey, M.T. (1977) Preliminaries to a theory of action with reference to vision.

In: Shaw, R. and Bransford, J. (eds) *Perceiving, Acting and Knowing: towards an ecological psychology*. Erlbaum, Hillsdale, NJ, pp.211–265.

Turvey, M.T (1990) Coordination. *Am Psychol* **45**, 938–953.

Utley, A., Steenbergen, B. and Astill, S. (2007) Ball catching in children with developmental coordination disorder: control of the degrees of freedom. *Dev Med Child Neurol* **49**, 34–38.

van Emmerik, R.E.A. and Newell, K.M. (1990) The effects of task constraints on intralimb and pen kinematics in a drawing task. *Acta Psychol* **73**, 171–190.

Vereijken, B. (1991) *The Dynamics of Skill Acquisition*. PhD thesis, Free University, Amsterdam.

Vereijken, B., van Emmerik, R.E.A., Whiting, H.T.A. and Newell, K.M. (1992) Free(z)ing degrees of freedom in skill acquisition. *J Mot Behav* **24**, 133–142.

Vereijken, B., Whiting, H.T.A. and Beek, P. (1992) A dynamical systems approach towards skill acquisition. *Q J Exp Psychol* **45A**, 441–471.

Section J

MEMORY

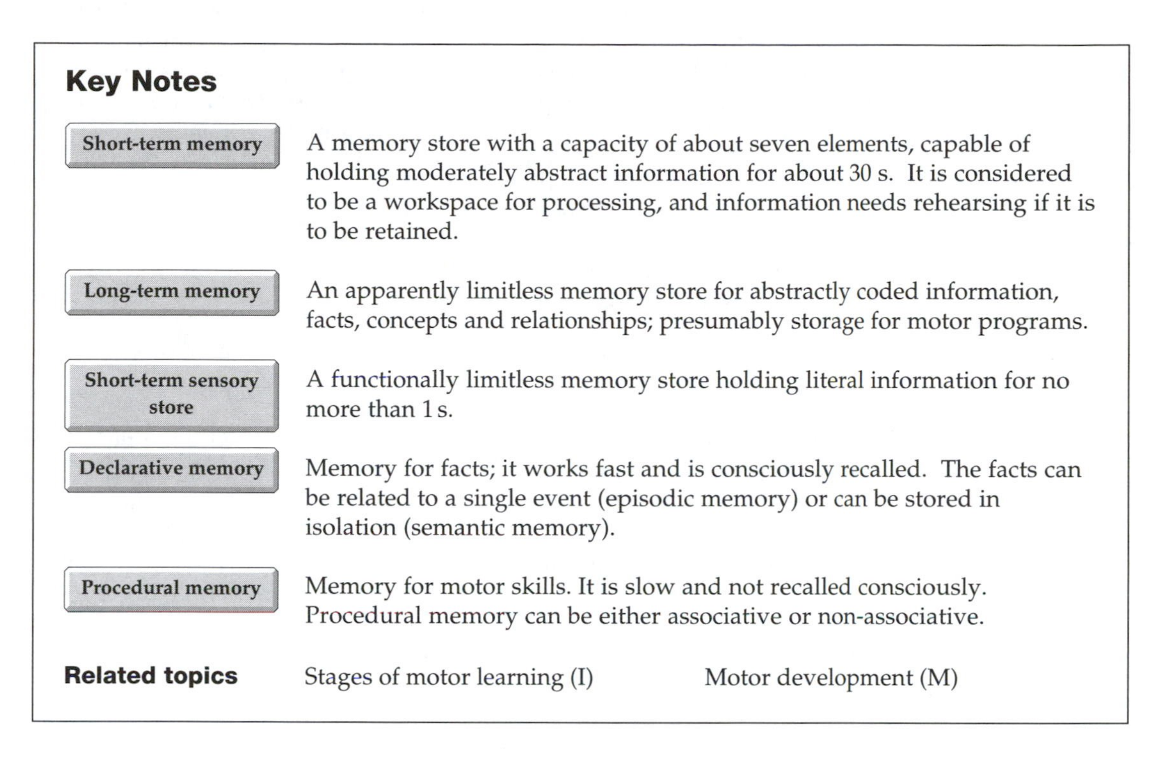

Key Notes

Short-term memory	A memory store with a capacity of about seven elements, capable of holding moderately abstract information for about 30 s. It is considered to be a workspace for processing, and information needs rehearsing if it is to be retained.
Long-term memory	An apparently limitless memory store for abstractly coded information, facts, concepts and relationships; presumably storage for motor programs.
Short-term sensory store	A functionally limitless memory store holding literal information for no more than 1 s.
Declarative memory	Memory for facts; it works fast and is consciously recalled. The facts can be related to a single event (episodic memory) or can be stored in isolation (semantic memory).
Procedural memory	Memory for motor skills. It is slow and not recalled consciously. Procedural memory can be either associative or non-associative.

Related topics Stages of motor learning (I) Motor development (M)

Memory plays a vital part in all our daily activities. Every day we have to remember facts such as numbers, names, and other factual information. Most of us can remember lines from plays or poems that we read many years ago. We also have an amazing capacity for remembering a vast array of information which only comes into its own at the pub quiz! Memory is therefore concerned with events that have just happened and events in the past. We are also able to remember procedures such as how to change a tire on a cycle, how to use a DVD, or how to make a particular cocktail. One remarkable aspect of memory is our ability to remember sequences of movement or motor skills. If you learnt to drive a car, learnt to ski, or to ride a bike five years ago and since then have not performed these tasks, you can still do it. Our memory for motor skills is therefore very good. If this were not the case we might be in the position of having to relearn to walk each morning. Memory forms an important part of motor control and it is closely related to learning. Many researchers have been interested in memory. The focus has been in three main areas: our memory capacity and storage, retrieval and retention, and loss of memory. This section provides an overview of models of memory, and discusses issues of retention and storage.

> My father started me skiing when I was 4 years old, and I skied every winter until I was 15, whereupon I stopped completely for 14 years. When I returned after this retention interval it seems safe to say that I was skiing as well as I ever did, and by the end of the day I was skiing better. (Schmidt, 1988)

So far we have focused on what happens to information as it enters the system, how it is processed and then acted upon to elicit a response. However, for motor skill learning to occur, information needs to be retained and then used again for repetition and refinement of a particular skill. The storage of this information and the altered behavior which occurs over time is memory. Our memory is the 'glue' that holds our intellectual processes such as perception, attention, decision making and problem solving together. Tulving (1985) considered memory to be capacity which enables us to use past experiences.

Memory can be divided into three stages:

1. *encoding*: the transmission of information into a form retainable in memory
2. *storage*: the holding of information
3. *retrieval*: the accessing of information from storage after a time interval and the use of the information to guide behavior.

The distinction between the stages is important but at the same time they are intimately connected and cannot be understood in isolation.

Atkinson and Shiffrin's multistore model

Perhaps the most widely accepted theory of memory is based upon the 'stage theory' proposed by Atkinson and Shiffrin (1968). The focus of this model is on how information is stored in the memory; the model proposes that information is processed and stored in three stages, comprising what Tulving (1985) would consider to be the declarative memory. In this theory, information is thought to be processed in a serial and discontinuous manner as it moves from one stage to the next. Atkinson and Shiffrin (1968) described the memory system in terms of a multistore model comprising three types of memory store: the short-term sensory storage (STSS), short-term memory (STM) and the long-term memory (LTM). How these three types of memory are linked and on a simple level how information is retained or lost is depicted in *Fig. 1*.

Short-term sensory storage (STSS)

All the information coming in from the environment, e.g. the movement of your opponent or the path of the rugby ball can be stored in the short-term sensory store (STSS). This lasts about a second or less. This is the first 'compartment' of memory, (*Fig. 2*). According to Atkinson and Shiffrin each sense has its own sensory store; the most important sensory stores are the auditory (echoic) and the visual (iconic), we also have a haptic memory for touch. The iconic memory can hold information for about 50 ms while the echoic memory store holds information for about 2 s (Sperling, 1960). Thus there is limited time for the processing of the information and only information that is important enters the short-term memory; the information that is unimportant is 'filtered out'. This process is called selective attention. Consequently, it is absolutely critical that the learner attends to the information, and this is more likely to happen if the information has an interesting feature and if the stimulus activates a known pattern.

Short-term memory

Only information that is attended to is moved from short-term sensory store (STSS) to the short-term memory (STM). Thus in sport it is important that attention is maintained while performing a skill, e.g. keep your eye on the rugby ball until you have caught it before trying to dodge your opponent. The attended information is then transferred to the short-term memory, which passes

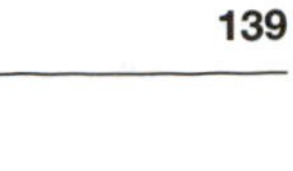

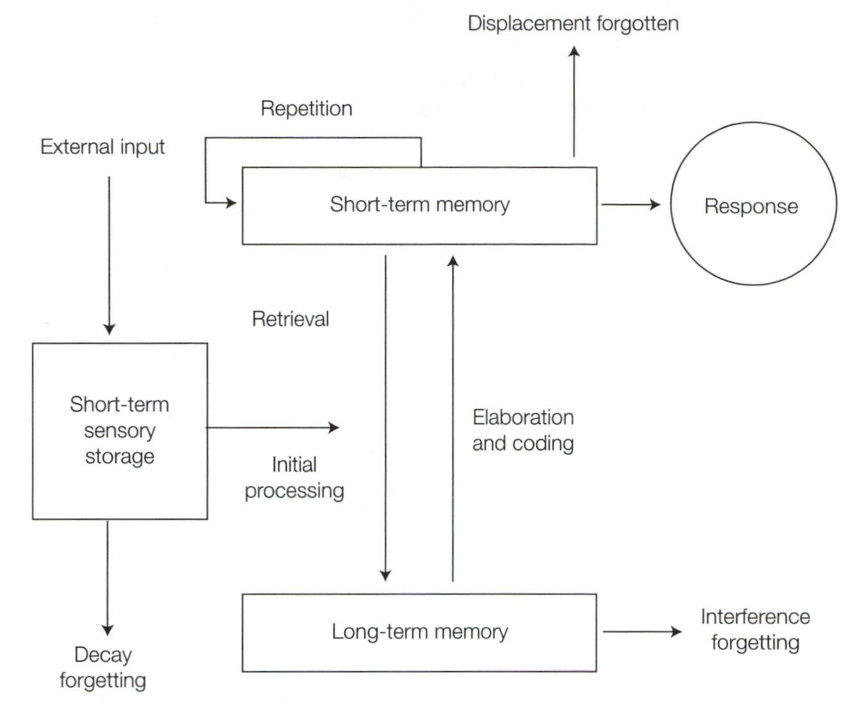

Fig. 1. The memory process controlling learning of a new skill.

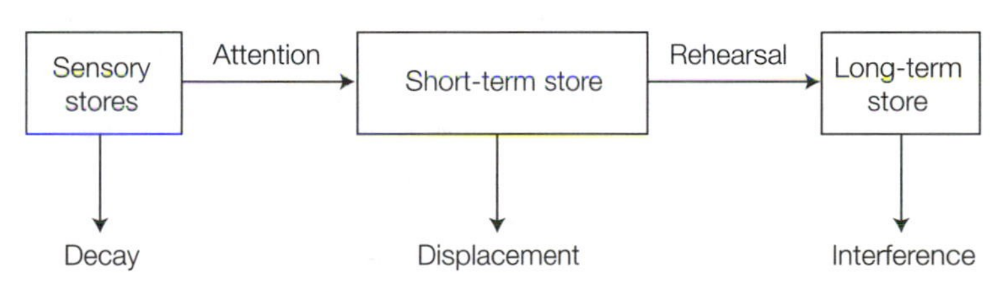

Fig. 2. The multistore model of memory.

rehearsed information to the long-term memory. The short-term memory has two key characteristics:

- *a very limited capacity* – only about 7 ± 2 digits/words can be remembered
- *fragility of storage*: any distraction usually causes forgetting.

The memory span for short-term memory is very short and holds information for a limited duration only, normally less than 60 s. Although everyone is different, generally the short-term memory can carry between five and nine separate items of information. However, this can be improved by chunking (Miller, 1956). For example, the numbers 5, 7, 7, 1, 9, 8, 6, 1, 0 could be chunked to 577, 198, 610. Now only three bits of information have to be retained. Miller (1956) found that healthy participants normally show a greater recall of material at the beginning and end of a list. For example, read these words aloud to a friend and then ask them to repeat them back to you and note which words they remembered:

Ball, Coach, Whistle, Bat, Field, Goal, Crowd

In free recall of this list, your friend is more likely to remember the first few items (ball, coach) and last few items (goal, crowd) compared to those in the middle of the list. These phenomena are accounted for by the primacy effect and recency effect respectively (Atkinson and Shiffrin, 1968). It is argued that the primacy effect happens because the person is able to rehearse the first few words and put them in the long-term memory (LTM). In addition, the recall of the words or digits is also influenced by the interval between words; short intervals = less ability for recall, long interval = greater ability for recall. Thus, the longer the interval the greater the amount of time you have for rehearsal. The recency effect occurs as the information is still active within the STM. This has important implications for the coach. As we now know, rehearsal must occur before the short-term memory can pass information to the long-term memory. However, due to its limited capacity, information is often lost before the learner can use it. To avoid this, the instructions a coach gives should be brief, succinct and given when the learner is paying attention. In addition, it may be better if the more important points are placed at the beginning or the end of the list of instructions. Other research has shown that our capacity for storing information is in fact constant at about seven items (Cowan *et al.*, 2004).

Box 1. Problems with the multistore model of memory

Atkinson and Shiffrin's (1968) multistore model has proved useful in showing the differences between the different memory stores and it is often used as a basis for modern theories; however, there are some disadvantages to this theory. The major problem is that it is too simple. STM cannot just be a unitary store (where all information is treated in the same way) because patients can have problems just with auditory-verbal aspects, but not other sounds (Shallice and Warrington, 1974). Similarly, LTM has been found to contain visual and verbal components which have different properties. The multistore model also overplays the role of rehearsal in obtaining information from STM for LTM.

Many experiments and researchers proved that STM is not a fixed storage place, but rather a mechanism for processing items or information. Baddeley and Hitch (1974) argued that the concept of the STM should be replaced with working memory. The working memory system consists of three components:

1. *central executive*: proposed as the control system of the model, responsible for strategy selection, planning, monitoring task performance and coordinating the other two components of the working memory
2. *phonological loop*: responsible for manipulating and maintaining speech-based information within the working memory
3. *visuospatial sketchpad*: responsible for the generation, manipulation, and retention of visual images.

To retain information in working memory, continuous rehearsing has to occur with the old and new information. This has to be done in such a way that we do not have to deal with too much information rehearsing at a time, as otherwise we would forget the old information and only remember the new. Baddeley argued that humans have a phonological loop and a visuospatial sketchpad which are generally regarded as two rehearsal systems.

The phonological loop is responsible for rehearsing and hence remembering speech-based information. In general, words of a shorter duration (i.e. take less time to rehearse) are remembered more than long words, e.g. bat, ball vs. player, referee. This suggests that the capacity of the phonological loop is determined by the temporal duration, and that memory span is determined by the rate of rehearsal. Furthermore, the phonological loop is vunerable to what is known as the phonological similarity effect (Conrad and Hull, 1964; Baddeley, 1996). Recall of words depends on how phonetically similar or how alike they sound. For example, recall of words that are phonologically similar, e.g. TBVCE, are harder to remember compared to letters that are phonologically dissimilar, e.g. TQRLP.

Another rehearsing system is the visuospatial sketchpad, which is used to rehearse visual images. This is used in the temporary storage and manipulation of spatial and visual information and so is very useful in everyday life. More recently, Logie (1995) has proposed that the visuospatial sketchpad can be further subdivided into the visual cache, which stores information about form and color, and the inner scribe, which deals with spatial and movement information.

Long-term memory

The LTM is where all information that enters the STM that is rehearsed, is stored. This memory has more capacity than the short-term memory, it is our day-to-day memory. The LTM is a store of well-learned past experiences and is used to compare against new experiences. It also stores responses used in these different situations. For example, a badminton shot is played by your opponent, which is very similar to one you have returned before. The STM links this information to the LTM, memory where a successful movement pattern is stored. The player can then play the shot using this information.

Box 2. Atkinson and Shiffrin (1968) multistore view of memory

1. *Sensory storage system*: large capacity, short memory duration
2. *Short-term storage system*: small capacity, medium memory duration
3. *Long-term storage*: large capacity, long memory duration.

Storage in the LTM

A number of researchers have proposed that memory should be categorized according to what information is stored there. In 1985, Tulving proposed three types of memory: episodic, semantic and procedural. However, in the last couple of decades there has been a tendency to describe two types of memory: declarative and procedural (Cohen and Squire, 1980; Anderson, 1987).

Declarative memory is memory that can be declared or expressed (explicit) and assists us in knowing 'what to do'. Declarative memory is all we have experienced in the form of information gained from childhood onwards. It is because of our declarative memory that we remember facts and events, such as, birthdays, telephone numbers or the rules of a game. Based on the work of Tulving (1985) there are two divisions of the declarative memory – episodic and semantic.

Episodic memory is an autobiographical memory for events or episodes that occur in a given time and place. It is memory of how, when and where something happened. For example, most people will remember exactly where they were when the World Trade Center tragically collapsed on 11 September in 2001. Semantic memory is memory for meaningful facts. Your name, math skills, and the definition of words are examples of semantic memory. Semantic memory is not tied to time and place and can thus be considered context free. For example, you remember that there are 24 hours in the day, but you probably don't remember where or when you learned this.

Procedural memory is memory storage of skills and procedures. This type of memory has also been referred to as 'implicit'. Procedural memory is involved in tasks such as remembering how to play the piano, drive a car, ride a bike or play football. This is 'know how' memory, it often can only be expressed by performing the specific skill, and people have problems verbalizing what they are doing and why. Procedural memory is therefore very important in human motor performance; however, it is also slow and, it needs many trials – or in other words a lot of rehearsal or practice. Once established, procedural memories are not lost even after many years without rehearsal. Cavaco *et al.* (2004) conducted a study that looked at the distinction between declarative and procedural memory. They ask patients with amnesia, and matched controls, to perform tasks that were based on real-world procedures, and declarative memory tasks. The patients with amnesia maintained performance and retained performance at the same level as the controls on the procedural tasks but they were much worse on the declarative tasks.

Tulving (2002) would argue that episodic memory grows out of semantic memory and so shares many of its features but also possesses its own features. He also suggests that this is the reason that in many amnesiac patients an impaired episodic memory is apparent but there is no impairment of their semantic memory. However, the majority of amnesic patients have impaired explicit memory (episodic + semantic memory) but intact implicit memory, so we need to be cautious before concluding that episodic and semantic memory form separate memory systems.

So far research has focused on the differences between episodic and semantic memory; we know very little about the similarities and interconnections between them. However, many, if not all, motor skills have both declarative and procedural memory components. Take the triple jump for example; this requires declarative memory for the rules about take-off, and procedural memory for the sequencing of the hop, skip and jump.

Not everything that is held in STM is eventually stored in LTM. The main factors that determine what is stored are rehearsal and meaning, and although it is possible to learn meaningless material, learning time is much shorter for meaningful material.

Ebbinghaus (1885) found that the likelihood of long-term recall is directly proportional to the total period of exposure to the material. He called this the *total time hypothesis*. Put simply, this hypothesis states that repeated exposure or consideration of a learning situation increases the likelihood that it will be stored in LTM. In terms of motor skill learning, this suggests that the more we practice setting the volleyball or dribbling the hockey ball, the more likely we are to remember how to do it and, more importantly, when to do it depending upon the game situation. However, it has also been suggested that it is not the total amount of rehearsal or exposure that determines recall. Baddeley (1986) found brief but regular exposure produced long-term recall more effectively than infrequent, prolonged exposure. This is known as the *distribution of practice* effect.

Research Highlight: Corkin (1968)
HM was a young man who had severe epilepsy. He had part of his forebrain removed to help control his condition. After the operation he had significant memory impairment. He could not learn new facts; he was unable to recognize people who visited him each day. As a result it was believed that the part of his brain that enabled the transformation from short-term memory to long-term permanent state was damaged. However, this theory was questioned when it became evident that he could learn new perceptual motor skills. He was given a mirror task in which one hand traces a complex task while viewing the task through a mirror (see figure below). The task is hard at first but generally people improve with practice. It was found that HM improved at the normal rate but each day when the task was presented to him he denied ever seeing it before.

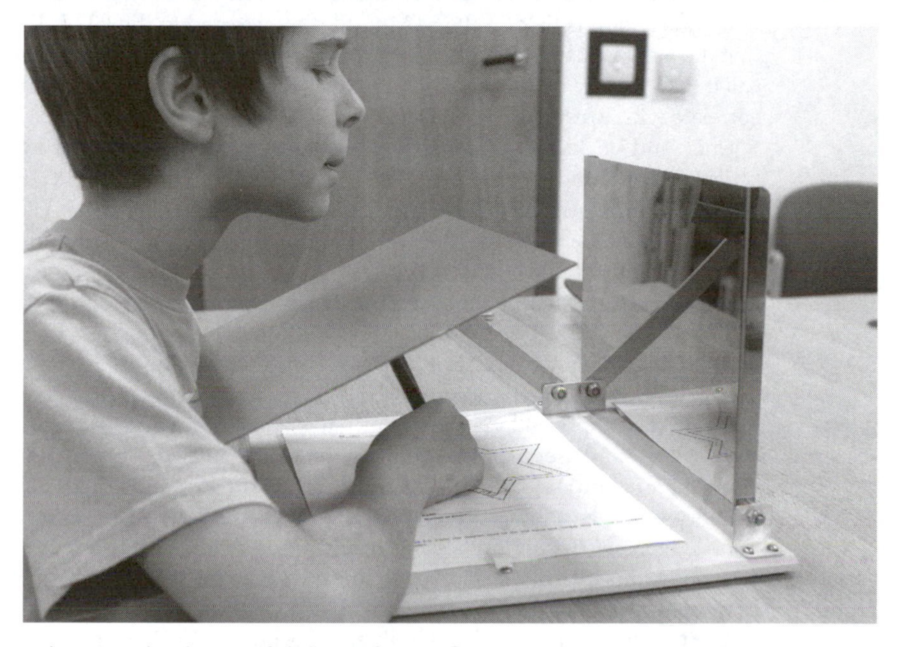

Participant demonstrating the use of similar equipment. Reproduced with permission from Corkin (1968).

Retrieval of information from the LTM
Information that has been stored in our memory has to be retrieved and acted upon to be of any use. It is generally agreed that there are two types of information retrieval from the LTM:

- *recall*: the recovery of information as a result of a conscious search
- *recognition*: the automatic recovery of information as a result of an external stimulus.

We are called upon to retrieve information from LTM almost every minute of every day. Our difficulty in retrieving information from LTM depends, in part, on how many cues we are given that would enable us to find the information.

One of the most difficult types of memory task involves remembering information with very few cues, or recall. In recall tasks, such as short-answer, fill-in-the-blank, and essay questions on exams, a question is asked, and we are required to produce the information. Another type of memory task simply requires us to pick out information that is the correct response to a question. This is called recognition. Common recognition tasks include multiple choice and matching questions on exams. Performance on recognition tasks is generally superior to performance on recall tasks, probably because recognition tasks give us so many more cues to find the information in LTM.

Craik and Lockhart's levels-of-processing model

Craik and Lockhart developed an alternative model to memory which attempted to address some of the problems with the multistore model of memory. Although many suggest that Craik and Lockhart (1972) advocate eliminating the distinction between the STM and LTM, the authors state that this is simply not the case (Lockhart and Craik, 1990). In fact they criticized the notion that information is held in the STM until it transferred to the LTM. They suggest that the STM is just a temporary activation of the LTM and actually preferred to call it the primary memory, which involves the activation of representations that correlate with present experience (Craik, 2002).

Craik and Lockhart (1972) see memory in terms of analyzing information in a series of stages and it is the build-up of analysis that contributes to the memory store, and proposed the levels-of-processing (LOP) model, the main contribution of which was to reinforce the idea of remembering as processing or as an activity of mind (Craik, 2002). The LOP model states that memory occurs on a continuum from shallow to deep, with no limit on the number of different levels. The shallow or superficial levels store information about physical or sensory characteristics, i.e. how it looks or sounds. The intermediate level of memory relates to recognition and labeling. The deep level is the storage of meaning (accomplished via semantic analysis) and networks of association. Both the shallow and deep processing involve retrieving pre-existing information knowledge about appearance, sound or meaning; and thus the resulting memory trace is a by-product of the processing involved in retrieving that knowledge, and deeper processing results in more elaborate, longer-lasting, and stronger memory traces. However, more recently Lockhart and Craik (1990) have acknowledged that this may not always be the case.

The rather general ideas of Craik and Lockhark (1972) were backed up by a series of experiments reported by Craik and Tulving (1975). One of these studies split participants into three groups, and gave each the same list of words. Although each group had the same list of words, they were all given different instructions:

- group 1 was asked to write down what the words looked like (e.g. bold, italic)
- group 2 was asked to write down words that rhymed with the words on the list (e.g. cat – mat)
- group 3 was asked to write down synonyms for each word on the list (e.g. cat – feline).

The participants were asked to write down as many of the words on the original lists as they could remember. Group 1 remembered very few, group 2 remembered approximately 50%, and group 3 remembered nearly all of those words on the list. In light of this work three levels of processing were suggested:

1. *structural processing*: shallow, surface processing – group 1 did not have to consider the words themselves

2. *phonetic processing*: deeper – group 2 had to imagine the words being spoken aloud in order to provide a rhyming word
3. *semantic processing*: processing to great depth – group 3 had to understand the meaning of the word to provide a synonym.

Craik and Lockhart also explored the concept of rehearsal, the process of cycling information through memory. Craik and Lockhart proposed two kinds of rehearsal. Type I processing (maintenance rehearsal) involves merely repeating the kind of analysis that has already been carried out. In contrast, type II rehearsal (elaborate rehearsal) involves a deeper, more-meaningful analysis of the stimulus. Elaboration is the process of adding more extensive information into the memory system. This serves to make existing information and incoming information more distinctive and unique, leading to better LTM than maintenance rehearsal. Craik and Lockhart suggested that only elaborate or the type II rehearsal should lead to an improvement in memory. As with other models Craik and Lockhart's LOP model was considered to have some disadvantages or problems. Firstly, there was no way to measure the depth to which information was processed, and some researchers (Baddeley, 1978) had difficulty with the model's focus on encoding rather than retrieval.

Forgetting

Any theory of memory must explain why information is forgotten over time. There are two main hypotheses as to why we forget; these are that we lose information due to decay, or interference. Decay refers to the loss of memory over time simply due to the passage of time. Interference refers to the situation in which new information may displace or corrupt older information. *Table 1* provides an overview of types of forgetting.

Table 1. Types of forgetting

Type of forgetting	Definition	Explanation	Our example	Your example related to sport
Decay	We can no longer recall information from our memory because of disuse	There was once a clear memory, but it has faded away because I did not use the information	I cannot recall my family's phone number when I was five because my family hasn't lived at that house in over 20 years	
Retroactive interference	Previously learned information is lost because it is mixed up with new and somewhat similar information	New information interferes with older information	I have trouble recalling my old phone number, because I get it mixed up with my new number	
Proactive interference	Current (new) information is lost because it is mixed up with previously learned, similar information	Earlier information projects itself forward and interferes with what we try to learn next	I have trouble recalling my new phone number, because I get it mixed up with my old number	

- *Decay theory* (spontaneous decay) assumes that when something is learned, a memory trace is formed that decays spontaneously.
- *Interference theory* (interference) views forgetting as being due to competing responses between the criterion tasks and tasks that have been learned before or after the criterion tasks.

There are two different types of interference. When new information pushes old items out of memory this is called retroactive interference; this type of interference works backward (retro), ruining memory for items entered earlier. For example, changing your telephone number may cause you to forget your old number. This is known as retroactive interference. However, there may also be times when older information 'resurfaces' and becomes confused with newer information. For example, you may suddenly recall an old telephone number and confuse it with your new one. This is known as proactive inhibition. Old items clogging up STM prevent the accurate entry of new information through proactive interference; this type of interference works forward ('pro'), straining memory for items entered later.

Other interesting aspects of memory are 'tip-of-the-tongue phenomena', where the right retrieval cues can facilitate our ability to suddenly remember something such as a word, name, or number. This illustrates the difference between things we don't know, even if we thought all day, and things we are sure are stored in memory somewhere. The key is to press the right buttons in order to release the information from storage. In situations such as eyewitness testimony where it is very important that exact details are remembered, many different techniques have been tried to assist witnesses with their recall. Going back to the scene of the crime/accident can help by providing important memory cues. We all have to find methods of remembering that have contextual meaning, and this often assists recall.

Disorders of memory

Amnesias (loss of memory) due to brain damage show that the STM and LTM are separable. Usually amnesia affects the LTM but leaves the STM untouched; however, there are reported cases of patients who have an intact LTM but impaired STM; this suggests that even if the STM is compromised, information can still enter the LTM. There are two main types of amnesia of the LTM: anterogade and reterogade. Someone with anterogade amnesia will generally have a good memory for the past up until the time of brain injury but will have extreme difficulty remembering anything that has happened since then. Such a person may not even remember what they had for breakfast that day. Reterogade amnesia is a condition in which individuals have trouble remembering events in their life prior to brain injury. Not all the individual's memories are lost, it is usually worse for those events that happened just before the injury.

Activity: Memory tasks

Task 1

Look at the words below and identify those that begin with capital letters:

Bat house clap Piano Flag ship

Now cover the words up and on a separate piece of paper write down the six words that were presented to you.

Task 2
Now I want you to look at the words below and identify those words that rhyme with team.

bucket grass dream cream door bream

Again, cover these words up and on a clean sheet of paper write down all the words that you can remember.

Task 3
Finally, from the words below identify those that can fit into this sentence: 'I ate . . . . for my dinner last night'.

fish paper peas pizza metal phone

Now, cover these words up and write down as many of them as possible. Finally, count up the number of words you remembered for each task and complete the graph below.

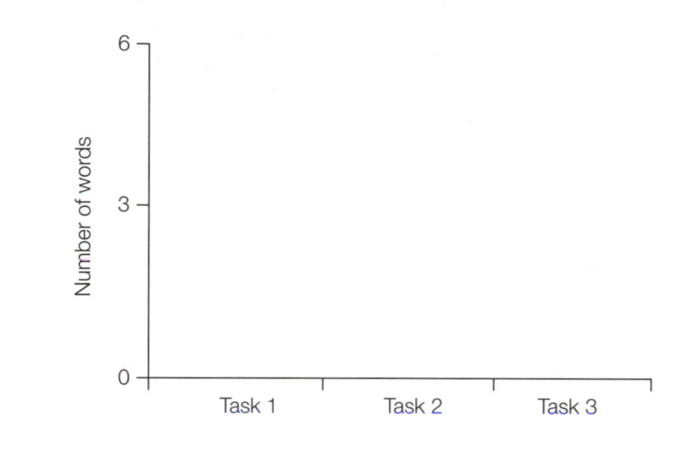

So what did you find? Like Craik and Lockhart you probably remembered more words from those presented in task 3 than in task 1 or 2, showing that the deeper the processing the more likely you are to remember something.

Enhancing memory

We are all concerned with enhancing memory, especially students who are about to take an examination in motor control. There are a number of strategies that can be used to enhance memory. Firstly, it is important to ensure that any learning situation is as meaningful as possible to the learner. One method of doing this is to use visual imagery of a movement, item or situation that has meaning to the learner. Researchers have also shown that giving a verbal label to a movement assists learning. For example, when learning a sequence, giving each movement a verbal label such as a number (counting) will assist learning. This has been proved in arm-positioning tasks (Shea, 1977) and positioning movements in children (Winter and Thomas, 1981). Another way to enhance learning is to divide that movement or the information to be learnt into parts. Here the learner is organizing the information. This is especially relevant to learning sequences of movements or complex skills. When employing this method the learner will break down complex skills into their component parts. As learning takes place, the learner will create less but larger component parts. A gymnast or ice dancer learning a new routine, or a slalom canoeist learning a complex route will do this. Eventually they will be able to perform the whole routine. The learner therefore creates structured organization or, put another way, places information in a group/order that has meaning. Structured information that can easily be organized appears to enhance learning (Millslagle, 2002).

Conclusion

Memory, and its functions and impact on learning and retention have been discussed in this section. Two components make up functional memory: working memory and long-term memory, and it is these that have created the greatest interest for researchers. Memory plays a key role in skill acquisition and in Section K, 'Implications for practice' we will revisit many of the issues raised in this section.

Further reading

Martell, S.G. and Vickers, J.N. (2004) Gaze characteristics of elite and near-elite athletes in ice hockey defensive tactics. *Hum Mov Sci* **22**, 689–712.

Schack, T. and Mechsner, F. (2006) Representation of motor skills in human long-term memory. *Neurosci Lett* **391**, 77–81.

Shadmehr, R. and Brashers-Krug, T. (1997) Functional stages in the formation of human long-term motor memory. *J Neurosci* **17**, 409–419.

Shea, J.B. and Wright, D. (1991) When forgetting benefits motor retention. *Res Q Exerc Sport* **62**, 293–301.

References

Anderson, J.R. (1987) Skill acquisition: compilation of weak-method problem solutions. *Psychol Rev* **94**, 192–210.

Atkinson, R.C. and Shiffrin, R.M. (1968) Human memory: a proposed system and its control processes. In Spence, K.W. and Spence, J.T. (eds) *The Psychology of Learning and Motivation: advances in research and theory*, vol. 2. Academic Press, New York, pp.89–195.

Baddeley, A.D. (1978) The trouble with levels: a re-examination of Craik and Lockhart's framework for memory. *Psychol Rev* **85**, 139–152.

Baddeley, A.D. (1986) *Working Memory*. Clarendon Press. Oxford.

Baddeley, A.D. (1996) Exploring the central executive. *Q J Exp Psychol* **49**, 5–28.

Baddeley, A. and Hitch, G.J. (1974) Working memory. In: Brower, G. (ed) *Recent Advances in Learning and Motivation*, vol 8. Academic Press, New York, pp. 47–90.

Cavaco, S., Anderson, S.W., Allen, J.S., Castro-Caldas, A.L. and Damasio, H. (2004) The scope of preserved procedural memory in amnesia. *Brain* **127**, 1853–1867.

Cohen, N.J. and Squire, L.R. (1980) Preserved learning and pattern-analyzing skill in amnesia: Dissociation of knowing how and knowing that. *Science* **210**, 207–210.

Conrad, R. and Hull, A.J. (1964) Information, acoustic confusion and memory span. *Br J Psychol* **55**, 429–432.

Corkin, S. (1968) Acquisition of motor skill after bilateral medial temporal-lobe excision. *Neuropsychologia* **6**, 255–265.

Cowan, N., Chen, Z. and Rounder, J.N. (2004) Constant capacity in an immediate recall-task: a logical sequel to Miller (1956). *Psychol Sci* **15**, 634–640.

Craik, F. (2002) Levels of processing: past, present . . . and future? *Memory* **10**, 305–318.

Craik, F. and Lockhart, R. (1972) Levels of processing: a framework for memory research. *J Verb Learn Verb Behav* **11**, 671–684.

Craik, F.I.M. and Tulving, E. (1975) Depth of processing and the retention of words in episodic memory. *J Exp Psychol Gen* **104**, 268–294.
Ebbinghaus, H. (1885) Über das gedächtnis: Intersuchungen zur experimentellen psychologie. Translated by Ruger H.A. and Bussenius C.E., 1913, and reissued by Dover Publications, New York, 1964.
Lockhart, R.S. and Craik, F.I.M. (1990) Levels of processing: a retrospective commentary on a framework for memory research. *Can J Psychol* **44**, 87–112.
Logie, R.H. (1995) *Visuo-spatial Working Memory*. Psychology Press, Hove.
Miller, G.A. (1956) The magical number seven, plus or minus two: some limits on our capacity for processing information. *Psychol Rev* **63**, 81–97.
Millslagle, D.G. (2002) Recognition accuracy by experienced men and women players of basketball. *Percept Mot Skills* **95**, 163–172.
Schmidt, R.A. (1988). *Motor Control and Learning: a behavioral emphasis*, 2nd Edn. Human Kinetics, Champaign, IL.
Shallice, T. and Warrington, E.K. (1974) The dissociation between short term retention of meaningful sounds and verbal material. *Neuropsychologia* **12**, 553–555.
Shea, J.B. (1977) Effects of labelling on motor short-term memory. *J Exp Psychol Hum Learn Mem* **3**, 92–99.
Sperling, G. (1960) The information available in brief visual presentations. *Psychol Monogr* **74**, 1–29.
Tulving, E. (1985) How many memory systems are there? *Am Psychol* **40**, 385–398.
Tulving, E. (2002) Episodic memory: from mind to brain. *Annu Rev Psychol* **53**, 1–25.
Winter, K.T. and Thomas, J.R. (1981) Developmental differences in children's labeling of movement. *J Mot Behav* **13**, 77–90.

Section K

IMPLICATIONS FOR PRACTICE

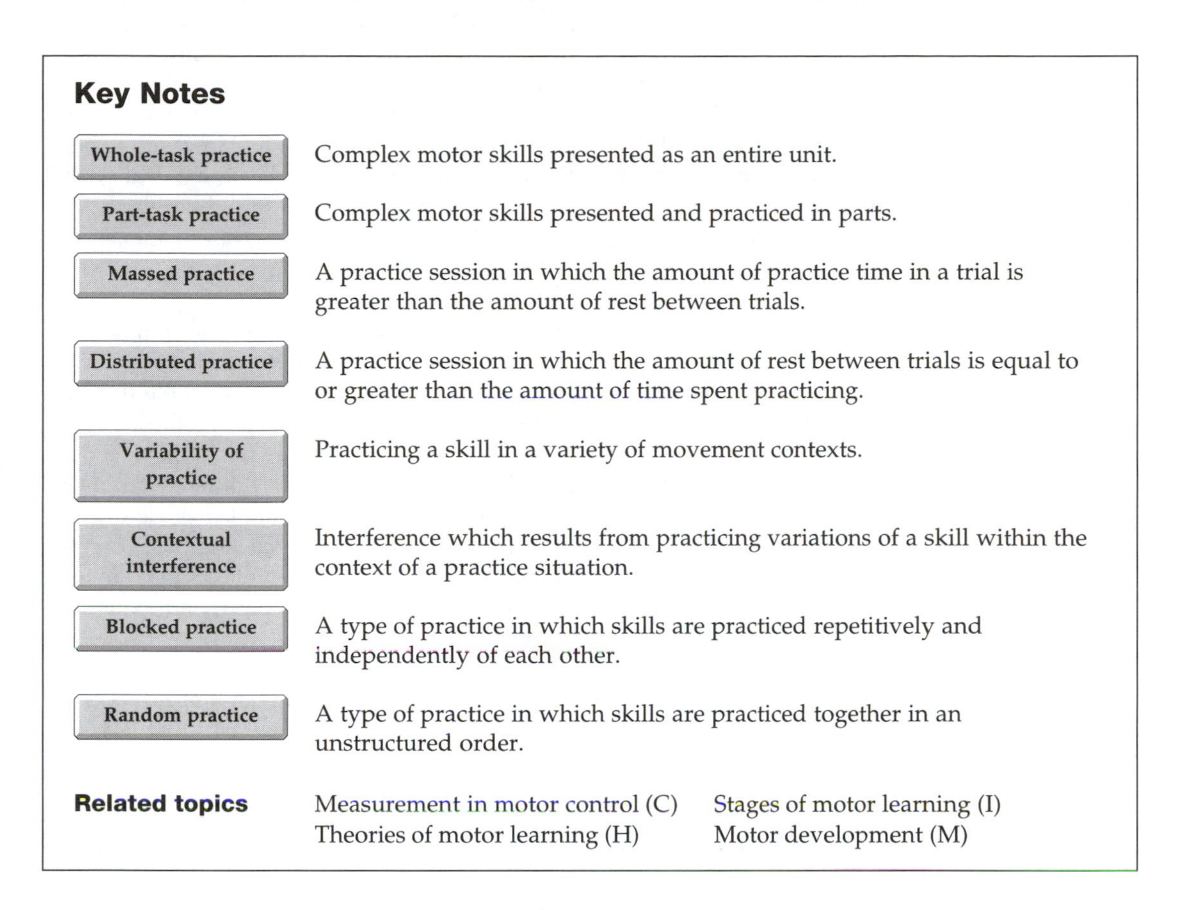

Key Notes

Whole-task practice	Complex motor skills presented as an entire unit.
Part-task practice	Complex motor skills presented and practiced in parts.
Massed practice	A practice session in which the amount of practice time in a trial is greater than the amount of rest between trials.
Distributed practice	A practice session in which the amount of rest between trials is equal to or greater than the amount of time spent practicing.
Variability of practice	Practicing a skill in a variety of movement contexts.
Contextual interference	Interference which results from practicing variations of a skill within the context of a practice situation.
Blocked practice	A type of practice in which skills are practiced repetitively and independently of each other.
Random practice	A type of practice in which skills are practiced together in an unstructured order.

Related topics

Measurement in motor control (C)
Theories of motor learning (H)
Stages of motor learning (I)
Motor development (M)

When coaching, teaching, or during rehabilitation we are generally attempting to improve performance. In previous sections we have considered theories of learning, stages of learning, how we control movements, and theories of control. All the information in these sections, and indeed in the ones that follow, have implications for coaching, teaching, and rehabilitation.

When learning a skill it must be remembered that practice in one context must often be transferred and repeated in another context. For example, the tennis player needs to perform successfully in practice matches and the competitive environment. In addition, performers have to successfully execute a variety of different actions; the tennis player has to be able to serve, volley, lob and return the ball using a forehand or backhand drive. All of these actions require a great deal of skill, and thus coaches have to be able to establish practice conditions that lead to the greatest probability of successful performances in all situations with all actions. Equally the patient relearning to walk may cope very well in the clinical setting but eventually they will have to walk in an environment that is much less controlled, such as the street. In light of this, a good coach, teacher or therapist will often ask themselves questions like: How should I structure my practice session to optimize learning? How should a particular skill be practiced? The following section discusses the learning environment and considers how skill learning can be maximized with particular reference to practice.

Whether we are talking about football, cricket, tennis or any other movement skill such as relearning to walk, all skilled performances have something in common; they have come about through effective practicing of the skill. In the world of movement science, we are always searching for practice methods that allow for optimal performance. A very young child quickly learns to kick a ball and make it go roughly in the direction he or she wants it to. However, to bend a ball around a wall, the way David Beckham can, requires a great deal of skill and practice. It is our job as coaches and/or therapists to find the right type of practice, and schedule practice so that it does not become too boring and so that each child, athlete or patient is able to give their 'best' performance.

Task analysis: breaking the task down into its component parts

One question that to date has been difficult for researchers to answer is whether it is better for the learner to practice the movement pattern in parts until each part has been perfected, or practice the movement pattern as a whole from the beginning. In the early 1960s Naylor and Briggs hypothesized that the organization and complexity characteristics of a skill could provide the basis for a decision to use whole or part practice. Task complexity (which is different from task difficulty) is directly related to task difficulty and/or the number of elements of the task, and subsequently the amount of information processing placed on the subject. For example, serving a tennis or volley ball are both tasks that are high in complexity, as is driving a car. Tasks that are low in complexity are skills such as throwing a dart or picking up a glass.

Organization was defined by Naylor and Briggs (1963) in terms of the interrelationship between the elements or parts of the task; more organized tasks are characterized by high interdependencies of the component. Tasks that rely very much on how one component is performed for the others to be successful have a high degree of organization. A good example of a task that has a high degree of organization is the triple jump. On the other hand, tasks that comprise many parts that are rather independent are considered to be low in organization, for example a dance routine.

Based upon Naylor and Briggs, assessing both the complexity of a task and the organization of a skill should help the coach or therapist to decide whether to use part or whole practice. The *whole method* involves learning the particular skill as a whole rather than trying to break it down into subroutines. In general, movement skills that are not too complex and that are highly organized in terms of their constituent parts are more suited to whole-task practice methods. For example, relatively simple skills such as catching a ball are best learned in this way. Many skills are very complex and even skills that might appear relatively simple, such as throwing the discus, will still appear very complex to novices. When teaching these, the coach should not present all aspects of the skill for practice. The learner would probably be overwhelmed by the quantity of information provided, and consequently would probably not be able to take in all of it. The *part method* is thought to be better for the more complex skills, or those that take longer to perform, e.g. a gymnastics routine. The whole skill is broken down into components that can be learned independently, and then they are recombined to form the whole skill. For example, footballers need to be able to run, dodge, jump, stop and turn, as well as kick a ball. They also need to learn how to control the ball with their feet, knees, chests, head and even their hands if they are a goalkeeper. These skills are generally performed as a combined whole while on the football pitch, but are often practiced on their own. Park *et al.* (2004) conducted an interesting study that examined the difference between whole and part practice of a 16-step movement sequence. Participants produced a right arm lever movement to target locations that were presented in sequence. The part group practiced the first eight sequences for one day and all 16 on the second day. The whole-practice group practiced all elements for both days. On the third day a retention and transfer task was performed, see *Fig. 1* for the results. The results showed no difference between part and whole practice in terms of learning the task. Part practice, however, resulted in better transfer in terms of performing the last eight elements (sequence B) but not for sequence A (the first eight elements). The authors state that this indicates that, compared to whole practice, part practice enables participants to eliminate parts of the movement without negatively influencing performance.

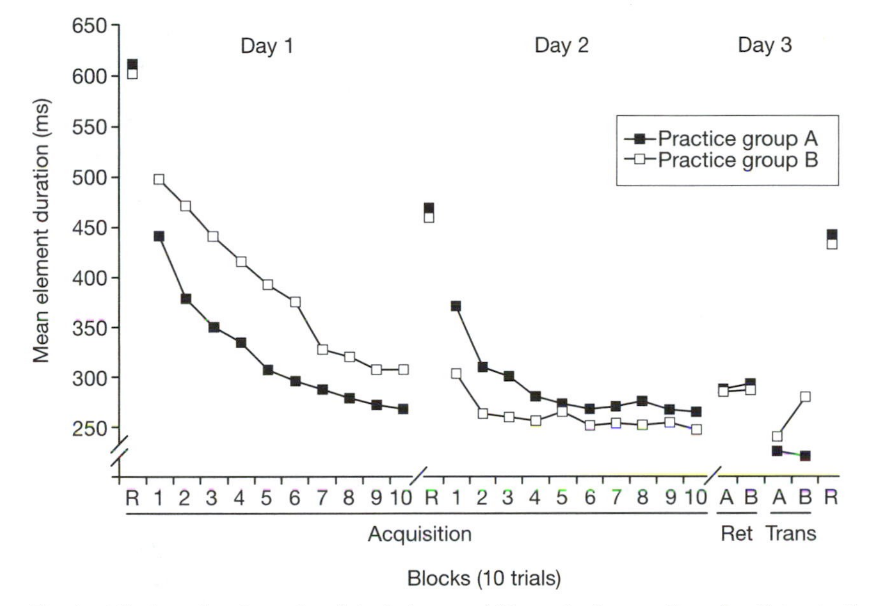

Fig. 1. Whole and part practice data during acquisition, retention, and transfer. Ret, retention; Trans, transfer. Adapted from Park et al. *(2004).*

When using part practice there is a need to devise subunits of skill, and decide how they should be practiced for maximum transfer to the whole skill. A starting point is to break the task to be learnt down into its component parts. This can take a number of forms and the task, context, and the level of expertise of the learner must be taken into account. Therefore, a number of facts have to be considered and evaluated. If, for example, we decide that the triple jump should be practiced in parts, there are still many questions to be answered. The performer can practice the hop, skip and jump separately; however, does the performer devote the same amount of time to the hop as the skip and the jump? Can the parts be practiced in the reverse order, e.g. the jump, the skip and then the hop? Does the coach recommend practicing the skip and jump together? In addition to this, what adaptations might be made to the context, can this be done in a gymnasium rather than the sand pit?

Wightman and Lintern (1985) suggested three methods of part practice that may be used depending upon the nature of the skill and the capabilities of the learner. These are simplification, segmentation and fraternization. *Segmentation* involves portioning a task into components along temporal or spatial dimensions; each of the parts is then combined with other parts, leading to the performance of the entire task. Prior to the practice of the task the size of the components is decided upon, depending upon the performer's capabilities. This is often called a *progressive-parts method*. Using this method, the triple jumper would learn the hop first, then skip, then the hop and the skip, then jump, then the hop, skip and jump. *Simplification* refers to the modification of movements to the basics of the skill within a particular environment.

The third method of part practice involves *fraternization* of the skill. This involves the dismantling of the coordination of a skill into several parts. In general, when a task is high in organization, a whole-practice approach would be better suited. However, skills can be often high in organization but benefit from a part-practice method (*Fig. 2*). For example, in skills where both arms are required to do something very different, it may be better to practice each arm separately before performing with the arms together.

There are a number of methods that can be employed to make the skill or parts of the skill less difficult and to enable learning to progress (*Box 1*). When breaking the task into its component parts, the teacher, coach or therapist can manipulate a number of aspects of the learning environment and this applies to both the sport and clinical setting.

Fig. 2. Skills can be broken down into component parts.

Box 1. Adaptations that can be made to assist learning. Adapted from Utley et al. *(2001)*

Adaptations to organizational arrangements
- Physical aspects of the classroom/gym/sports hall/pitch etc
- Organization of grouping within the year group, within the session
- Organization of support: before, during, and after the session.

Adaptations to teaching methods
- Use of a variety of teaching styles
- Extending their active learning/activity components
- Use of active learning strategies, self-evaluation
- Variable practice
- Attention to spoken explanations – clarity, pace, vocabulary
- Monitoring of perfomers' understanding, use of additional memory aids
- Regular feedback and meaningful praise
- Assessment and modification of written materials – readability, clarity, vocabulary
- Consistency
- Over-learning
- When applying sanctions ensuring that they do not have hidden rewards.

Adaptations to curricular tasks and activities
- Tailoring tasks to the performers' interests
- Modifications to equipment size, weight, height, color, texture, and sound
- Modifications to space (adjustments to size of space, use of cones to mark out zones, use of texture to mark out zones, use of walls to field the ball)
- Modifications to time (some performers may need more time to complete an activity, others should be stretched by reducing the time)
- Modifications to rules
- Use of a variety of resources and equipment
- Providing alternatives to or modifications of pupil responses.

Conditions of practice

> . . . practice, when properly undertaken, does not consist of repeating the means of solution of a motor problem time after time, but is the process of solving this problem again and again by techniques which we changed and perfected from repetition to repetition. . . . Practice is a particular type of repetition without repetition. (Bernstein, 1967)

After breaking down the skill as appropriate (*Fig. 2*), the next question is how the skill should be practiced. As coaches and therapists we need to decide how many activities should be practiced each session and how much time should be devoted to each activity. The belief that the more you practice the better might not be correct, and a balance has to be struck between quantity and quality. One of the major issues for the coach, teacher and therapist is how to space or distribute practice for optimal learning. *Massed practice* is where the skill is practiced repeatedly over a long period of time. Skills tend to be learned better this way, though fatigue and boredom can cause some problems. More experienced performers and athletes tend to respond better to this type of practice. *Distributed practice* is where the sessions are short, spread out over time with rest periods or changes of activity interspersed with the main practice. This is a good way to offset boredom, but there is a danger that the skills may not be acquired as well as with massed practice. This technique is probably best for younger players or those new to the sport. *Variable practice*, as the name suggests is a mixture of both types. Depending on the skill level and experience of the individual, one or other or both types of practice might be employed by the coach. For part of the performance massed practice might be best, but for others, distributed practice favored. Variety is always useful in training sessions as it will reduce tedium and keep people motivated. A number of research studies have been conducted to determine which practice schedule leads to better performance and learning. It must be remembered that performance and learning are different, and that retention and transfer also need to be considered (see Section H). When reviewing studies that look at the benefits of practice it must be remembered that there is conflicting evidence as to which type of practice is best.

Massed practice

During massed practice the time spent active or actually practicing the skill is greater than the time spent resting. Massed practice will involve fewer blocks of practice but the blocks will be more intensive and the performer will repeat the action many times within the block of practice. During massed practice the performer will have very short or no periods of rest. The benefits or negatives of massed practice will very much depend upon the nature of the skill to be practiced. For skills that are relatively discrete such as a tennis, short massed practice which involves short rest periods may work very well. A table tennis player, as seen in the *Fig. 3*, may well benefit from massed practice of a particular shot, such as a smash. However, for better learning it may well be best to spread practice over a number of days or weeks. Therefore massed practice may work well for practice in a particular session, but massed practice sessions over time may not be beneficial. Vearrier *et al.* (2005) found that intensive massed practice worked well for retraining balance post-stroke, therefore its is important to consider the individual, the task, and the context when planning practice.

The following points should be considered:

- for more continuous skills such as learning to ride a bike, practice should be short and with plenty of rest
- for more discrete skills such as a tennis shot, massed practice with short rests may be best
- more frequent learning sessions will assist learning and retention; practicing a skill once a month may not work well
- trying to rush skill learning by over-massing practice will not work well
- practice sessions should not be so long that fatigue becomes a factor
- There is some crossover in the use of the term massed and distributed practice, and the next section and Research Highlight are also relevant to the information on massed practice.

Distribution of practice

When planning practice, regardless of whether this is in the sporting, educational or rehabilitation context, the frequency of practice and the length of

Fig. 3. Massed practice of a forehand smash.

recovery has to be decided. Over the years there has been a range of research that has explored the distribution of practice. Researchers have considered how massed, distributed, random and blocked practice influence performance, learning, retention and transfer. We have reviewed a few studies here, but there are many and, as with other areas of research, the findings present a contrasting picture of what is best 'practice'. As you will see from the few studies reviewed here, one of the main causes of a lack of consensus is the range and variety of tasks, contexts and participants involved in these studies. This at times makes comparisons difficult. Studies have also manipulated rest periods so that the period of practice is longer than the period of rest. In distributed practice, however, the period of rest is often equal to or less than the work period. Bourne and Archer (1956) conducted a study looking at the benefits of rest when learning a rotary pursuit task. They concluded that the longer the rest period the better the performance. Other research has supported this work, with Baddeley and Longman (1978) examining the best way to train postal workers on a keyboard task. The aim was to achieve a typing speed of 80 key stokes per minute. The participants were divided into four groups with varying practice schedules. One group had one hour's training once a day for 12 weeks. Another had one hour's training twice a day for 6 weeks, a third two hours' training once a day for 6 weeks, and the final group two hours' training twice a day for 3 weeks. There was therefore a variety of ways in which practice took place, from more-distributed to more-massed practice. The results showed that distributed practice was by far the best, and after a 9-month retention test the group that had the most massed practice performed worse, with little difference between the other groups.

Other studies have found similar results to Baddeley and Longman, and distributed practice would appear to be better than massed practice. Savion

Fig. 4. Skills can be acquired using a variety of practice schedules.

Lemieux and Penhume (2005) looked at the effects of amount of practice and length of delay on the learning and retention of a timed motor sequence task. Participants were assigned to a varied-practice condition or a varied-delay condition. In the varied-practice condition, participants received either 1, 3, or 6 weeks of practice followed by a 4-week delay recall. In the varied-delay condition, participants received three blocks of practice followed by a varied delay of 3 days or 2, 4, or 8 weeks. The amount of practice did not affect learning or retention, but distribution of practice did affect learning and retention. Savion Lemieux and Penhume conclude that delay may well allow for consolidation of learning. Vearrier *et al.* (2005) conducted a study in which the purpose was to test if intensive massed-practice intervention (participants practiced 6 hours per day for 2 consecutive weeks) could significantly improve balance function post-stroke. The ability of the subjects to recover from a balance perturbation improved during training, and further still during the maintenance phase, ending 3 months post-intervention. Intensive massed practice of standard physical therapy produced significant results in balance retraining with patients post-stroke. Studies of this nature are of particular interest as they advance our knowledge of how best to rehabilitate patients post-injury in a variety of conditions.

Although there is conflicting evidence, it would appear that massed practice does not produce as positive results as distributed practice, especially in terms of retention and transfer. It may well be that massed practice causes fatigue and does not allow the learner time to evaluate, reflect or make appropriate

Research Highlight: Dail and Christina (2004)

An interesting study that examined putting in golf was conducted by Dail and Christina (2004). Participants included 90 novice golfers with a mean age of 22.3 years. They were randomly divided into two groups: a massed-practice and a distributed-practice group. The task was to attempt to hole a putt or get as close as possible from a distance of 370 cm. The number of correct putts or distance from the hole was recorded. Each group completed 240 acquisition putts, with the massed group completing them in one session. The distributed group practiced 60 putts per day for 4 days. The results indicated that the distributed-practice group outperformed the massed-practice group during both acquisition and retention (60 putts). The figure below shows the results, including the retention test which took place on three occasions (1 day, 7 days, and 28 days later (R1, R7, R28).

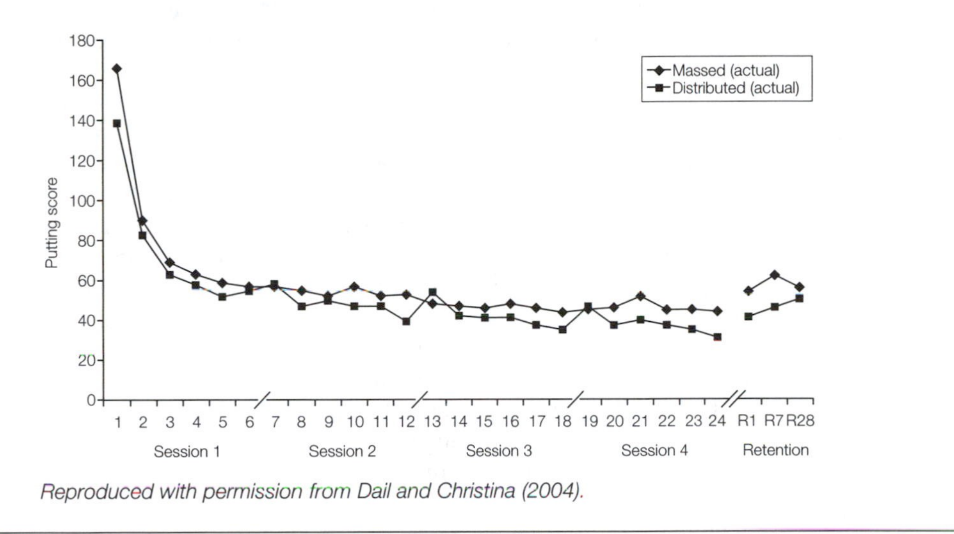

Reproduced with permission from Dail and Christina (2004).

cognitive associations. Another form of practice that is popular and that is well researched is variable practice.

Variable practice

A consistent characteristic of motor learning theories is that practice variability is beneficial to skill acquisition. For example, Schmidt's schema theory (Schmidt, 1975) states that practicing variations of a movement (variable practice) will be of benefit when the movement must later be recalled (retention) or when producing a novel form of movement (transfer). Schmidt's schema theory predicts that practicing a variety of different ways to perform a motor skill during a practice session provides the learner with an opportunity to apply different parameters to a movement skill and develop their generalized motor program (see Section H). Recall schema is therefore strengthened by variable practice. Thus, a goalkeeper may practice stopping the ball from different distances, different heights and different trajectories and different set pieces, such as a penalty kick, corner or indirect free kick. He or she may also use balls of different sizes or the goals may be adjusted in size (*Box 1*).

In Section I, Gentile's (2000) two-stage theory of motor learning was introduced, and it is generally agreed that two guiding principles emerge from this model. The first is that although Gentile (2000) emphasized the need during practice for the learner to experience a variety of regulatory and non-regulatory context conditions, she also states that variable practice should not really be introduced until the learner has established an appropriate coordination pattern for the task at hand. In light of this, consistent practice may be more beneficial for the learner at this stage. Secondly, the type of variability that should be introduced will differ with respect to the type of skill being learned and the environment in which it will be ultimately performed.

A number of researchers have considered how variability of practice influences learning and retention. Much of the research has been conducted to test predictions about schema theory. Over the years the range of tasks to be learnt has varied from discrete to continuous, and from few to many degrees of freedom to be controlled. Shea and Kohl (1991) looked at comparisons between variable-practice conditions in order to consider the influence this had on retention. Participants were split into three groups and trained on a task that involved being able to precisely exert a given force on a stress gauge. One group practiced only on the criterion force (criterion only) while another other group practiced at four different forces as well as the criterion force (criterion and variable). The final group practiced the criterion task and completed additional spaced practices of the criterion task (criterion–criterion). The results indicated (*Fig. 5*) that the groups that practiced exclusively at the target force initially acquired the skill faster. However, when tested later, the criterion-varied group had retained their skill much better compared to the other groups. Variability of practice therefore benefited skill retention.

From the 1970s to the present day, there has been a variety of studies examining variable practice and its influence on skill acquisition and retention. These studies cover a massive range of tasks, contexts, and individuals. McCracken and Stelmach (1977) found that when learning to knock over a barrier with the

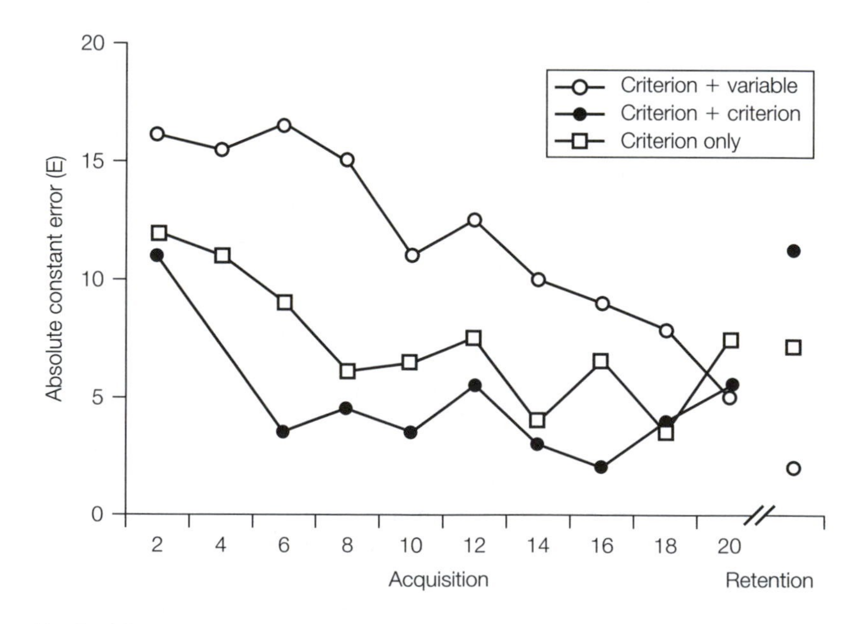

Fig. 5. The benefits of variable practice on retention. Adapted from Shea and Kohl (1991).

hand in a set time, the variable group outperformed the constant group. As with Shea and Kohl (1991) they found that during two retention periods the variable group outperformed the constant group even though they had not during the acquisition period. Recent studies have looked at more complex tasks. Granda-Vera and Montilla (2003) looked at practice schedule, acquisition, retention and transfer when 6-year-old children performed a throwing task. They found that participants in the variable practice group showed better performance than the blocked group in acquisition, retention and transfer. Douvis (2005) also found that variable practice produced better practice results when using variable practice in learning a forehand in tennis. Heitman *et al.* (2005) looked at the effect of specific versus variable practice on a pursuit rotor task. The variable group practiced at varying speeds and the specific practice group performed 30 trials at 45 rpm on day one, and on day two all groups performed 15 trials at 45 rpm and 15 trials at 75 rpm. It was found that the specific group had higher retention scores and the variable group had high transfer scores. The authors conclude that this reflects the influence of context on continuous motor skills.

Williams and Hodges (2005), in a paper examining practice and skill acquisition in football, presented an interesting diagram showing the relationship between variable practice, contextual interference, and practice activity (*Fig. 6*). It is this relationship that is of interest to coaches, teachers, and clinicians and in the next part of this section we will explore this.

Implementing practice variability: contextual interference

Having established that practice variability enhances learning, we must now further consider how a teacher, coach or therapist should organize the practice session. One way to do this which is particularly popular is to apply the phenomenon known as *contextual interference* (CI). CI is a functional interference

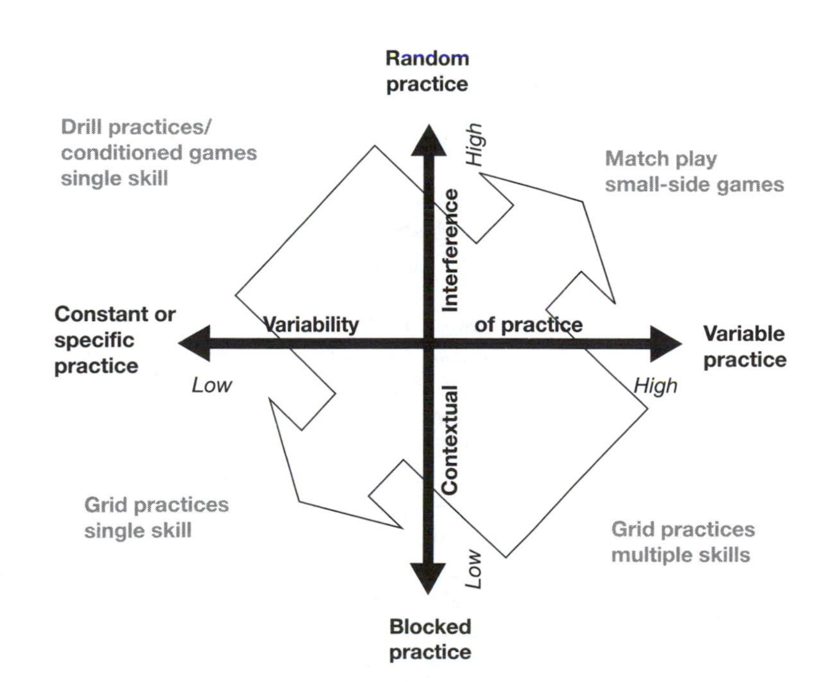

Fig. 6. The relationship between variable practice, contextual interference and practice activity in football. Reproduced with permission from Williams and Hodges (2005).

introduced in practice which results in several movements being practiced, and refers to the learning effectiveness of random vs. blocked practice.

Low CI results in the performer practicing only one skill in the practice session – *blocked practice*. For example, when learning to play hockey, the player may practice shooting for 10 minutes, passing and receiving the ball for 10 minutes and then spend their final 10 minutes practicing dribbling. Although blocked practice schedules assist individuals in initial decision making and performance enhancement, when interference is introduced to simulate real-world conditions, there is a substantial erosion of learning. In stark contrast, a high degree of CI can occur when a learner practices several different but related skills in the same practice session – *random practice*. If we consider the performer learning to play hockey again, in the same 30-minute session the players may practice shooting, passing and receiving and dribbling for five consecutive attempts, and the order in which the skills are practiced is changed throughout the session. Under conditions where the environment is dynamic and the individual is required to adapt to variations in the surroundings, random practice schedules are superior to blocked practice sessions for retention of motor skills. A third type of practice, *serial practice*, would see the player practicing dribbling, passing, receiving and the shooting in five consecutive trials, but with the order in which the skills are practiced kept the same throughout. These different kinds of practice are not unique to teaching in physical education, it is also characteristic of, for example, a therapy session in which the patient is learning to use their wheelchair, in that the patient will have to practice this skill on a variety of different surfaces.

Interest in CI can be traced back to Battig's (1966) work in the area of verbal learning. Battig (1966) proposed that 'intertask facilitation is produced by intratask interference' (p.227) (intertask facilitation being the capacity to transfer and perform between different tasks). This was in direct contrast to predominant methods of conceptualizing the role of interference in learning and memory at the time.

Shea and Morgan (1979) showed that the CI effect as observed by Battig was applicable to the motor skill acquisition in a study which saw participants move their arms as quickly as possible through three different three-segment patterns. In response to a stimulus light, the participants picked up a ball with their right hand and knocked over a series of freely moveable wooden barriers and then brought the ball to a final resting position. Some participants were instructed to use blocked practice (low CI) and some participants used random practice (high CI). Participants in the blocked-practice group practiced task A 18 times followed by task B and task C 18 times each. The random-practice group practiced each task 18 times but the order of presentation was random (e.g. ABCB-CABAC). Shea and Morgan's (1979) data showed that although the participants in the blocked-practice group had a better performance, they did not have better retention and transfer of the skill.

Since 1979, substantial laboratory-based research on CI has been conducted (Magill and Hall, 1990; Brady, 1998). However, several researchers have questioned the utility of laboratory-based research, claiming that it lacked ecological validity to real-world environments. Despite Del Rey's (1989) notion that the laboratory tasks that had been selected were chosen because their processing demands were similar to those required in sports, many of the tasks used to investigate the CI effect were simple and involved minimal degrees of freedom, which by their very nature resulted in rapid gains in skill. To validate the theoretical notion of CI, researchers began translating the concept into instructional procedures that applied to the practitioner, and tested them out with real-world sporting activities.

To date, researchers have found some evidence for the effect of contextual interference on the acquisition of sport skills. Goode and Magill (1986) followed by Wrisberg and Lui (1991), showed that practicing badminton serves with high or low CI did not elicit a difference between the two groups in acquisition, but that during retention and transfer differences between the two groups were apparent. This, however, was limited to the performance on the short serve. In addition, these studies did not demonstrate as Battig predicted, that blocked practice would cause superior performance during acquisition.

Strong support for the CI effect has been demonstrated by Hall *et al.* (1994). These workers randomly assigned 30 baseball players to either a control group or a group that performed blocked or random practice. The random and blocked groups received two additional batting-practice sessions each week for 6 weeks, while the control group received no additional practice. The extra sessions consisted of 45 pitches, 15 fastballs, 15 curveballs, and 15 change-up pitches. The blocked group received all 15 of one type, then 15 of the next type, and finally 15 of the last type of pitch, in a blocked fashion. In contrast the random group received these pitches in a random order. Pretest analysis showed no significant differences between the groups. However, when the transfer test results were examined the random group performed with reliably higher scores than the blocked group, who performed better than the control group. When comparing the pretest to the random transfer test, the random group improved 56.7%, the blocked group 24.8%, and the control group only 6.2%, superbly demonstrating the CI effect.

Smith and Davies (1995) examined the effect of contextual interference on the learning of the Pawlata roll in a kayak. This study was motivated by the fact that at the time many instructors in Britain were advocating learning the Pawlata roll in one direction only to a criterion of accuracy, thereafter transferring to the opposite direction. However, Smith and Davies (1995) suggest that skill retention would be better served by practicing on alternate sides and therefore proving contextual interference. In light of this, Smith and Davies conducted a study in which 16 undergraduate students with no kayaking experience were randomly allocated to either a low-CI group or a high-CI group, which practiced the skill on alternate sides. Analysis of the data showed that the high CI group was quicker to achieve successful performance in retention (full roll) and transfer (half roll) tests, regardless of the direction of the roll, one week later. In addition, and perhaps more surprisingly, the high-CI group also acquired the skill more quickly than the blocked-practice group. Overall, these data showed the recommended training methods at the time for learning the Pawata roll were not appropriate.

There is remarkable consistency in the motor learning literature favoring high CI. However, there is limited research that has examined the effect of high CI on the relearning of activities of daily living. Given that the therapist often has limited time with a patient, as well as the fact that there is a limited number of tasks and task variations that can be experienced in a given rehabilitation setting, practice sessions that promote retention of the gains that are made with the therapist and that facilitate transfer to other contexts away from the clinic are highly desirable.

Weeks *et al.* (2003) examined the role of variability in practice structure when learning to use upper-extremity prosthesis. The purpose of the study was to examine the acquisition, retention, and transfer of selected prehension skills under conditions of low and high CI to establish the most effective practice context for patients. As an adequately sized population of recent upper-extremity amputees was not available for the study, the training was given to able-bodied participants who wore a simulator that mimicked actual upper-extremity prosthesis in form and function. Forty-eight participants were randomly distributed into two groups for skill-acquisition training: those who practiced three different prehension tasks with the simulator in a random practice order or those who practiced the three tasks with the simulator under a blocked-practice order. During acquisition, the groups practiced the three tasks on two consecutive days. On the third day, two different tests of learning were administered: a retention test on the tasks practiced in acquisition, and an intertask transfer test on three tasks similar to those practiced in acquisition. Analyses of data showed that both the random and blocked groups showed significant improvements in initiation time and movement time to perform each task across the two days of acquisition. Thus, regardless of degree of 'randomness' inherent in the schedule, practice did indeed promote functional use of the prosthesis. In addition, both practice schedules were equally effective for promoting skill retention. However, during the intertask transfer phase of the study, the random-acquisition group demonstrated significantly greater proficiency in performing the new tasks than the blocked-acquisition group. Weeks *et al.* (2003) suggest that their data show that persons learning to use upper-extremity prostheses may be better able to transfer skill to new prehension tasks by practicing under random-practice conditions. Further studies like that performed by Weeks *et al.* (2003) are required to validate the concept of CI for rehabilitation, and this will add to the ever-increasing pool of resources for therapists to draw on while practicing evidence-based medicine.

The influence of expertise, task complexity, and contextual interference was examined by Ollis *et al.* (2005). Novice and experienced knot tiers took part in the study in which they practiced tying knots that were simple and complex.

After establishing baseline-knot tying abilities the novice group and the expert group practiced in three experimental conditions, low CI, moderate CI, and high CI. The results of this experiment can be seen in *Fig. 7*. It was found that high levels of CI scheduling enhance learning for novices even when practicing a complex task, and also that CI benefits are more apparent as the learners engage in tasks with high transfer distality. The authors conclude that complexity and experience influence the CI training effect in motor situations that have high contextual meaning.

Theoretical explanations

Despite the fact that it seems that the CI effect is equally applicable in both the laboratory and the applied setting, there is still considerable debate surrounding the accepted explanation for the CI effect. Current theoretical explanations include the *elaboration hypothesis* (Shea and Zimmy, 1983; 1988) and the *action-plan reconstruction hypothesis* (Lee and Magill, 1985). However, both rely heavily on the role of cognitive effort or the more effortful processing engendered by random practice schedules; and the deficient or decreased processing resulting from blocked schedule (Albaret and Thon, 1998; Brady, 1998; Immink and Wright, 1998).

The elaboration hypothesis (Shea and Zimmy, 1983; 1988) is based upon the levels-of-processing framework of memory (see Section J) and relies heavily on

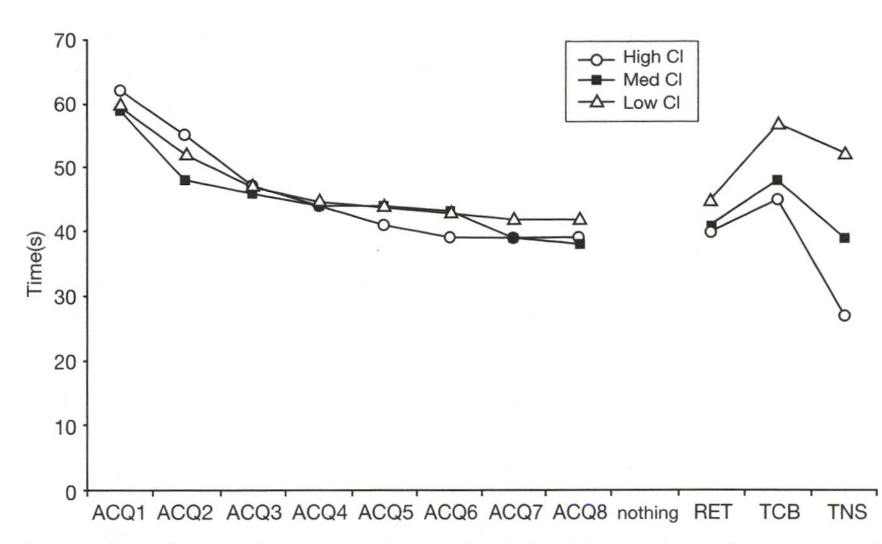

Fig. 7. Time taken to tie (a) simple-knots group (novice) and (b) complex-knot group (experts) in varying contextual-interference situations. ACQ, acquisition; RET, retention; TCB, transfer complex and blindfold; TNS, transfer normal and sighted. Reproduced with permission from Ollis et al. *(2005).*

the notion of intratask and intertask processing and their use by the learner during practice. In blocked practice, learners primarily use intratask processing or within-task analysis, since only one task resides in the working memory. However, random practice encourages the learner to engage in both intra- and intertask processing since multiple tasks reside in the working memory. Shea and Morgan (1979) suggested that random practice works by actively engaging the learner in the learning process, by forcing them to try out a number of different strategies to complete the motor task. The majority of these skill variations are held in the working memory (see Section J), allowing the performer to compare each variation, to see what is similar or dissimilar and what works or does not work. Consequently, there is further elaboration of the memory representations of the skill variations that a leaner is practicing. In the end these variations becomes distinct from each other and they can then be used when the environmental and task constraints dictate it. Therefore, according to the elaboration view of CI, random practice leads to a more enriched representation, whereas a more impoverished encoding results under blocked conditions.

Lee and Magill (1985) proposed an alternative but not contradictory explanation for the CI effect – the action plan reconstruction hypothesis. Lee and Magill (1985) claimed that the elaboration hypothesis could not adequately account for the random group's acquisition performance deficits. According to Schneider *et al.* (1995), the retention benefit of random practice can be attributed to the extensive retrieval practice that occurs as a result of a continual interchange of task information in the working memory (see Section J) from trial to trial. Essentially, when learners are engaged in random practice they are required to regenerate an action plan, based upon information stored in the long-term memory store, on the next practice trial for that skill variation. This regeneration is thought to have occurred because the learner has partially or completely forgotten the action plan that she or he developed on the previous practice trial for that skill variation (Lee and Magill, 1985). In contrast, the performer who engages in blocked practice can use the action plan that she or he developed on the previous trial.

The contextual interference effect: is there an exception to the rule?

An overview of CI research typically shows that acquisition performance is improved with repetitive blocked practice, while retention and transfer performance, perhaps the most important indicators of training effectiveness (Salas and Canon-Bowers, 2001), benefit from random-practice scheduling (Brady, 1998). Since the original demonstration of the CI effect in the motor domain (Shea and Morgan, 1979), this effect has been observed using a variety of subject populations (Del-Rey *et al.*, 1982; Edwards *et al.* 1986; Ollis *et al.* 2005; Keller *et al.* 2006) and environmental settings (Goode and Magill, 1986; Shea and Wright, 1991). Although it appears that there is a large amount of evidence that demonstrates the effectiveness of the CI effect, the data also show that it does not always apply to all motor learning situations (for a good review see Magill and Hall, 1990; Wulf and Shea, 2002). Brady (1998) cites many important factors that can influence the CI effect, such as amount of practice, cognitive effort, intrinsic interest and/or motivational processes, trait anxiety and/or self-efficacy. Some of these are discussed below.

Task-related characteristics

Magill and Hall (1990) suggested that the primary limiting factor for the CI effect was the task's characteristics. They noted that skills that require many variations that are controlled by different generalized motor programs would be affected more by high CI. More specifically, they suggested that practicing tasks controlled by different motor programs rather than the same one resulted in more interference and consequently more effortful processing. In contrast, tasks that are governed just by one motor program require more parameter (force or timing) modification. Unfortunately, this proposal has not been well supported by applied research, indeed, a near opposite trend is apparent with much of the data suggesting that high CI enhances learning tasks with similar movements.

Practicing more complex tasks under high interference conditions may prove too difficult for early learners to elicit CI effects. Research which has investigated the effects of CI on the learning of volleyball skills has produced equivocal results (French *et al.* 1990; Bortoli *et al.*, 1992) and Magill and Hall (1990) proposed that high CI may overwhelm novices in the early stages of skill acquisition and that learners need to be somewhat proficient before they can take advantage of random practice. This proposition was later supported by Hebert *et al.* (1996). Alternatively, more expert individuals may not find simple skills challenging enough even under random conditions. Research has shown that in general experienced learners benefit more than novices from high levels of CI scheduling (Guadagnoli and Lee, 2004; Magill and Hall, 1990) and that as task complexity increases, the relative benefit from increasing levels of CI should decrease (Albaret and Thon, 1998; Wulf and Shea, 2002).

Learner characteristics

Although predominantly an under-researched area, it has been suggested that learner characteristics (e.g. personality) may interact or mediate with the CI effect. For example, an individual's self-efficacy may dictate whether low or high CI is more appropriate for skill acquisition. Bandura (1997) coined the phrase 'self-efficacy' as the belief that an individual has in his or her own ability to execute a certain course of action or obtain a specific outcome. As Brady (1998) notes, it seems likely that highly efficacious individuals would adapt more readily to high-CI conditions. In contrast, as blocked practice accelerates acquisition, individuals with low efficacy may benefit more from this type of practice, as performance accomplishments are a potent source of efficacy (Bandura, 1997).

In line with Gentile (2000), research has suggested that children may be better suited to blocked rather than random practice. At first high CI may be more difficult for beginners and children, and blocked practice is more likely to help the learner establish the basic movement pattern. Jarus and Goverover (1999) investigated the varying conditions of contextual interference within 5–11 year olds who practiced the task of throwing beanbags under either low contextual interference (blocked practice), high contextual interference (random practice), or medium contextual interference (combined practice). Analysis indicated that only the 7-year-old subjects differed in their performance in the various practice groups. On the other hand, the 5 and 11 year olds did not differ in performance in the different practice groups. This suggested that developmental age may interact or mediate the effects of CI. Granda-Vera and Montilla (2003) also investigated the effects of CI in young children aged six years. Participants on a variable-practice schedule showed better performances than those on a blocked schedule, and the difference between the two groups was significant in the acquisition, retention and transfer phases. These results support the hypothesis that contextual interference enhances skill learning.

Conclusion

The role of practice in motor learning and retention remains an area of great interest for all those involved in designing practice. As we constantly strive to improve performance, improve rehabilitation and ensure we are more efficient, there will always be debate about what is the best to practice a skill. What is clear is that we must always consider the nature of the task, the individual, and the context in which the skill is both learnt and performed.

References

Albaret, J.M. and Thon, B. (1998) Differential effects of task complexity on contextual interference in a drawing task. *Acta Psychol* **100**, 9–24.

Baddeley, A.D. and Longman, D.J.A (1978) The influence of length and frequency of training session on the rate of learning to type. *Ergonomics* **21**, 627–635.

Bandura, A. (1977) Self-efficacy: towards a unifying theory of behavioural change. *Psychol Rev* **84**, 191–215.

Battig, W.F. (1966) Facilitation and interference. In: Bilodeau, E.A. (ed) *Acquisition of Skill*. Academic Press, New York, pp.215–244.

Bernstein, N. (1967) *The Coordination and Regulation of Movements*. Pergamon, Oxford.

Bortoli, L., Robazza, C., Durigon, V. and Carra, C. (1992) Effects of contextual interference on learning technical sport skills. *Percept Mot Skills* **75**, 555–562.

Bourne, L.E. and Archer, E.J. (1956) Time continuously on target as a function of distribution of practice. *J Exp Psychol* **51**, 25–33.

Brady, F. (1998) A theoretical and empirical review of the contextual and interference effect and the learning of motor skills. *Quest* **50**, 266–293.

Dail, T.K. and Christina, R.W. (2004) Distribution of practice and metacognition in learning and long-term retention. *Res Q Exerc Sport* **75**, 148–155.

Del-Rey, P. (1989) Training and contextual interference effects on memory and transfer. *Res Q Exerc Sport* **60**, 342–347.

Del-Rey, P., Wughalter, E. and Whitehurst, M. (1982) The effects of contextual interference on females with varied experience in open sport skills. *Res Q Exerc Sport* **53**, 108–115.

Douvis, S.J. (2005) Variable practice in learning the forehand drive in tennis. *Percept Mot Skills* **101**, 531–545.

Edwards, J.M., Elliot, D. and Lee, T.D. (1986) Contextual interference effects during skill acquisition and transfer in Down's syndrome adolescents. *Adapt Phys Act Q* **3**, 250–258.

French, K.E., Rink, J.E. and Werner, P.H. (1990) Effects of contextual interference on retention of three volleyball skills. *Percept Mot Skills* **71**, 179–186.

Gentile, A.M. (2000) Skill acquisition: action, movement, and neuromotor processes. In: Carr, J.H. and Shephard, R.B. (eds) *Movement Science: foundations for physical therapy*, 2nd Edn. Aspen, Rockville, MD, pp.111–187.

Goode, S. and Magill, R.A. (1986) Contextual interference effects in learning three badminton serves. *Res Q Exerc Sport* **57**, 308–314.

Granda-Vera, J. and Montilla, M.M. (2003) Practice schedule and acquisition, retention, and transfer of a throwing task in 6-yr-old children. *Percept Mot Skills* **96**, 1015–1024.

Guadagnoli, M.A. and Lee, T.D. (2004) Challenge point: a framework for conceptualizing the effects of various practice conditions in motor learning. *J Mot Behav* **36**, 212–224.

Hall, R.G., Dominques, D.A. and Cavazo, R. (1994) Contextual interference effect with skilled baseball players. *Percept Mot Skills* **78**, 835–841.

Hebert, E.P., Landin, D. and Solomon, M.A. (1996) Practice schedule effects on the performance and learning of low and high skilled students: an applied study. *Res Q Exerc Sport* **67**, 52–58.

Heitman, R.J, Pugh, S.F., Kovaleski, J.E., Norell, P.M. and Vicory, J.R (2005) Effects of specific versus variable practice on the retention and transfer of a continuous motor skill. *Percept Mot Skills* **100**, 1107–1113.

Jarus, T. and Goverover, Y. (1999) Effects of contextual interference and age on acquisition, retention and transfer of a motor skill. *Percept Mot Skills* **88**, 437–447.

Keller, G.J., Li, Y., Weiss, L.W. and Relyea, G.E. (2006) Contextual interference effect on acquisition and retention of pistol shooting skills. *Percept Mot Skills* **103**, 241–252.

Lee, T.D. and Magill, R.A. (1985) Can forgetting facilitate skill acquisition? In: Goodman, D., Willberg, R. and Franks, I.M. (eds) *Differing Perspectives in Motor Learning and Control*. North Holland, Amsterdam, pp.3–22.

Magill, R.A. and Hall, K.G. (1990) A review of contextual interference effects in motor skill acquisition. *Hum Mov Sci* **9**, 349–367.

McCracken, H.D. and Stelmach, G.E. (1977) A test of the schema theory of discrete motor actions. *J Mot Behav* **11**, 193–201.

Naylor, J. and Briggs, G. (1963) Effects of task complexity and task organization on the relative efficiency of part and whole training methods. *J Exp Psychol* **65**, 217–244.

Ollis, S., Button, C. and Fairweather, M. (2005) The influence of professional expertise and task complexity upon the potency of the contextual interference effect. *Acta Psychol* **118**, 229–244.

Park, J.H., Wilde, H. and Shea, C.H. (2004) Part-whole practice of movement sequences. *J Mot Behav* **36**, 51–61.

Salas, E. and Canon-Bowers, J.A. (2001) The science of training: a decade of progress. *Annu Rev Psychol* **52**, 471–499.

Savion Lemieux, T. and Penhume, V.B. (2005) The effects of practice and delay on motor skill learning and retention. *Brain Res Exp Brain Res* **161**, 423–431.

Schmidt, R.A. (1975) A schema theory of discrete motor skills learning. *Psychol Rev* **82**, 255–260.

Schneider, V.I., Healy, A.F., Ericsson, K.A. and Bourne, L.E. Jr. (1995) The effects of contextual interference on the acquisition and retention of logical rules. In: Healy, A.F. and Bourne, L.E. Jr. (eds) *Learning and Memory of Knowledge and Skills: durability and specificity*. Sage, Thousand Oaks, CA, pp.95–131.

Shea, C.H. and Kohl, R.M. (1991) Composition of practice: influences on the retention of motor skills. *Res Q Exerc Sport* **62**, 187–195.

Shea, J.B. and Morgan, R.B. (1979) Contextual interference effects on the acquisition, retention and transfer of a motor skill. *J Exp Psychol Hum Learn Mem* **5**, 179–187.

Shea, J.B. and Wright, D.L. (1991) When forgetting benefits motor retention. *Res Q Exerc Sport* **62**, 293–301.

Shea, J.B. and Zimmy, S.T. (1983) Context effects in memory and learning movement information. In: Magill, R.A. (ed) *Memory and Control of Action*. Elsevier, Amsterdam, pp.145–166.

Shea, J.B. and Zimmy, S.T. (1988) Knowledge incorporation in motor representation. In: Meijer, O.J. and Roth, K. (eds) *Complex Movement Behaviour: the motor-action controversy*. Elsevier, Amsterdam, pp.289–314.

Smith, P.J. and Davies, M. (1995) Applying contextual interference to the Pawlata roll. *J Sports Sci* **13**, 455–462.

Utley, A., Whitelaw, S.A. and Hills, L.A. (2001) Equality, differentiation, support and a whole school approach – key factors in inclusive physical education for children with special educational needs. *Br J Teach Phys Educ* **Autumn**, 37–41.

Vearrier, L.A., Langan, J., Shumway-Cook, A. and Woollacott, M. (2005) An intensive massed practice approach to retraining balance post-stroke. *Gait Posture* **22**, 154–163.

Weeks, D.L., Anderson, D.I. and Wallace, S.A. (2003) The role of variability in practice structure when learning to use an upper-extremity prosthesis. *J Prosth Orthot* **15**, 84–92.

Wightman, D.C. and Lintern, G. (1985) Part-task training strategies for tracking and manual control. *Hum Factors* **27**, 267–283.

Williams, A.M. and Hodges, N.J. (2005) Practice, instruction and skill acquisition in soccer: challenging tradition. *J Sports Sci* **23**, 637–650.

Wrisberg, C.A. and Lui, Z. (1991) The effect of contextual variety on the practice, retention and transfer of an applied motor skill. *Res Q Exerc Sport* **62**, 406–412.

Wulf, G. and Shea, C.H. (2002) Principles derived from the study of simple skills do not generalize to complex skill learning. *Psychonom Bull Rev* **9**, 185–211.

THE ROLE AND FUNCTION OF FEEDBACK

Key Notes

Augmented feedback

Information about the performance of a skill which is added to sensory feedback (intrinsic feedback) and comes from a source external to the performer, e.g. a coach or therapist.

Knowledge of results

A type of augmented feedback given after completion of the movement that gives the performer information about the movement outcome or performance.

Knowledge of performance

A type of augmented feedback that gives information about the movement characteristics that led to the movement outcome or performance.

Related topics

Theories of motor learning (H)
Implications for practice (K)
Motor development (M)

Imagine you are teaching someone how to swim or to play tennis, or that you are working with a patient in a clinic who is relearning to walk after a stroke. In both of these situations, the people practicing these skills will make a lot of mistakes and would probably benefit from receiving information about how they are doing, what is right or wrong or what they should try next time. This information is called feedback and it provides the various clues needed to solve the problems posed when learning or relearning a motor skill. In Section G we considered the internal sensory feedback mechanisms (such as vision, proprioception and audition) that tell the performer about various aspects of the movement, such as how it felt or how accurate they were. In this section we consider the role the educator, coach or therapist plays in providing movement-related information that supplements the feedback that is readily available from internal sensory sources (intrinsic feedback). This supplementary type of information is called extrinsic or augmented feedback, and although after reading Section K we know how important it is to the motor learning process, there are still many questions to be answered such as how should you give feedback? Should you give qualitative or quantitative feedback? Should you give feedback after every trial? These questions relate to fundamental issues confronting both the physical education teacher, the coach of an Olympic gymnastic squad and the physiotherapist, and the aim of this section is to go some way to answering these questions.

Augmented feedback: what is it, what does it do and do we really need it?

As we can see from *Fig. 1*, there are two categories of feedback. The first is intrinsic or sensory feedback. This type of feedback is available during and/or after a person performs a skill and is discussed in more detail in Section G. The second form of feedback that we consider in this section is extrinsic or augmented feedback. Augmented refers to adding to or enhancing intrinsic feedback with an external source. The external source may be a device such as a timer, video recorder, biofeedback system or the teacher, coach or therapist. Examples of augmented feedback being used are readily available in the sporting arena. For example, a coach may tell the hurdler whether the trailing leg was brought forward too late even though the performer could feel this for him or herself and heard the hurdle fall down as they hit it. In a clinical setting, biofeedback may be used to help patients with hemiparesis to re-establish symmetrical postural alignment (Shumway-Cook *et al.*, 1988). *Figure 2* shows a diagram of the role of augmented feedback in physiotherapy adapted from Hartveld and Hegarty (1996), who elaborate on the importance of such feedback in aiding the physiotherapist during rehabilitation.

Augmented feedback plays many roles in the skill learning process. It in part answers questions about their performance, such as: How did I do? What shall I do differently next time? It allows the performer to get closer and closer to the most energetically efficient and successful way to complete a task no matter how simple or complex. Feedback helps the learner search the perceptual-motor workspace for the potential solution to their movement problem. Feedback also directs attention. You may recall from Section E how in any movement situation there are many stimuli that a performer may use to complete a movement task. We also discussed the importance of selectively attending to one or a few of these stimuli that are essential to selecting the right movement response. A coach may direct the performer's attention to one of these stimuli to enhance the chance of the performer selecting the correct response. Augmented feedback also motivates the performer to continue striving for success. Motivation is a

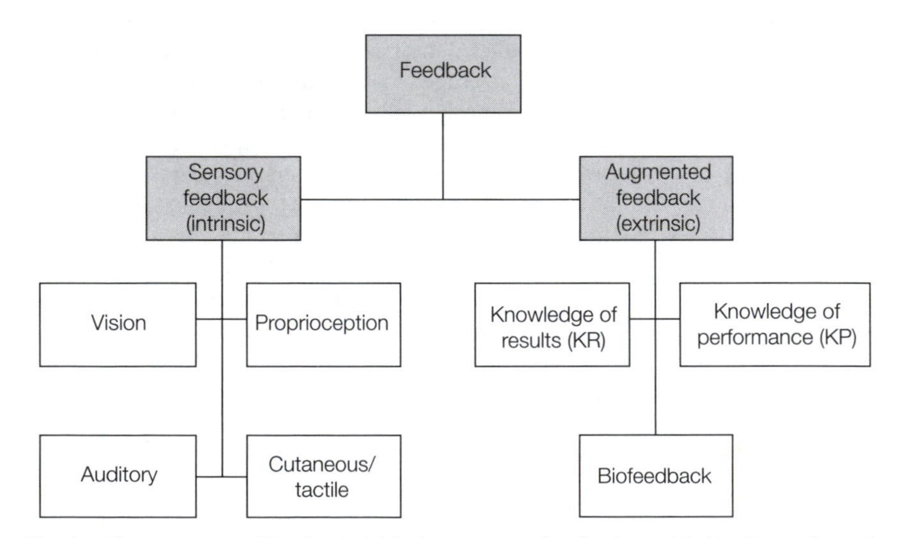

Fig. 1. The two forms of feedback: intrinsic or sensory feedback provided by the performer's receptors (see Section G) and extrinsic or augmented feedback provided by an external source.

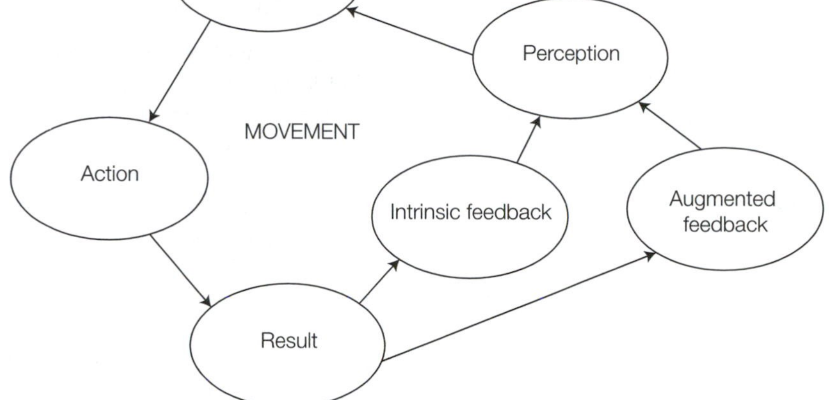

Fig. 2. The interaction between action and perception with feedback as a key component. Adapted from Hartveld and Hegarty (1996).

particularly important organismic constraint on successful movement and is beyond the scope of the present book.

As we discussed in Section G, feedback is especially important for skill learning and development, in fact, along with practice it is considered to be one of the most potent variables that affects motor skill learning (Newell, 1991). However, in some skill-performance situations, performers do not have access to the critical sensory information that they can then use to determine what they need to do to improve. Thus for some, augmented feedback is of particular importance or can even be considered critical to motor skill learning. Consider, for example, a situation in which people do not have the available sensory pathways needed to detect intrinsic feedback arising from the various sensory sources. In light of the organismic constraints placed upon these performers due to injury or disease, augmented feedback has to act as a substitute for this intrinsic or sensory information and now becomes essential to skill learning. Augmented feedback is also important for the performer who does not have access to visual information due to task constraints, a hammer thrower for example. During the turns the performer cannot see the area into which the hammer needs to be thrown (i.e. the target). Until the performer is better able to interpret the proprioceptive information arising from this movement, augmented feedback is essential for skill learning.

So now we know what function augmented feedback plays in motor learning, that it can be used in the sporting arena or clinical setting and that it is important for the motor skill acquisition process. However, there are still many questions that need answering, such as:

1. what kind of feedback should we give?
2. should we provide feedback during the movement, immediately following completion of the movement or at some time after the movement has been completed?

3. should the augmented feedback describe the outcome of the movement and/or the movement pattern itself?
4. how much feedback should be provided?
5. how often should feedback be provided?

The remainder of the section will expand on these issues with the ultimate aim of showing you what kind of feedback you should employ, when to use it and what it should tell the performer about their performance, and finally how it can be used to optimize the skill-learning process.

Types of augmented feedback

Perhaps the first important distinction that needs to be made concerning augmented feedback is the difference between knowledge of results (KR) and knowledge of performance (KP) (*Table 1*). Knowledge of results (KR) is supplementary information that is given to the performer after completion of the movement and describes the outcome of the movement in terms of the movement goal. For example, the teacher may tell the student in hockey practice 'the shot was one meter wide at the left post'. Here the teacher has given the student information about the performance outcome. However, KR is only helpful when it provides information performers are not able to obtain on their own, if for example they are not skilled enough to use intrinsic feedback or they cannot see the result of their performance.

Table 1. Comparison of KR (knowledge of results) and KP (knowledge of performance)

Knowledge of results (KR)	Knowledge of performance (KP)
Used to confirm own assessment of action	More useful when skill requires a series of specific movements (e.g. a serial skill)
Needed when intrinsic information is not available	Useful for more complex, multiple-degree-of-freedom movements and thus more applicable to real-world situations
Motivates the learner	It helps with the correction of performance errors, i.e. helps the learner achieve the goal more quickly
Promotes active motor skill learning	Motivates and reinforces learning

Newell (1974) illustrated how important augmented feedback is in a situation in which the sensory or intrinsic feedback needed to perform a skill is not available or an individual is not capable of using it. Participants had to make a 24 cm lever movement in 150 ms. Although they could see their arms, the lever and the target, the results showed that their success at learning the movement depended on how many times (out of 75 trials) they had received KR about the accuracy of their responses. These results indicated that in the initial learning stage, those participating did not have a good internal referent for determining what 150 ms was. They needed augmented feedback for at least 52 trials to establish such a referent; this information was used to perform the movement without the need for augmented feedback.

Research Highlight: Wierinck *et al.* (2006)

An interesting study by Wierinck *et al.* (2006) looked at manual dexterity training in dental students (a task that we all hope they learn to perform quickly and painlessly). Thirty-six first-year dental students took part in the study. The task involved a cavity preparation activity during which augmented feedback was given continuously (100%) or at a reduced rate (66% of the time), while performing the task. A control group received no augmented feedback. All the students were given a preparation score produced by Dentsim™. This system scores students from 0–100 by comparing the operator's performance to an ideally prepared tooth stored in the systems memory. Preparation scores were taken pre-test, on five occasions during acquisition, one day post-training and 4 months post-training. In addition, two transfer tests were also performed 1 day and 4 months post-training. As can be seen in the figure below, the preparation scores for both feedback groups were better than for the control group (CO), with a significant difference in the final acquisition trial (S5) and the transfer trial (T2). Therefore there is little difference on retention and transfer when giving 66% (66% FB) or 100% feedback (100% FB). In terms of preparation time, the 66% feedback group outperformed both the 100% feedback group and the control group during acquisition, and during retention and transfer there was a significant difference from the control group but not between the feedback groups.

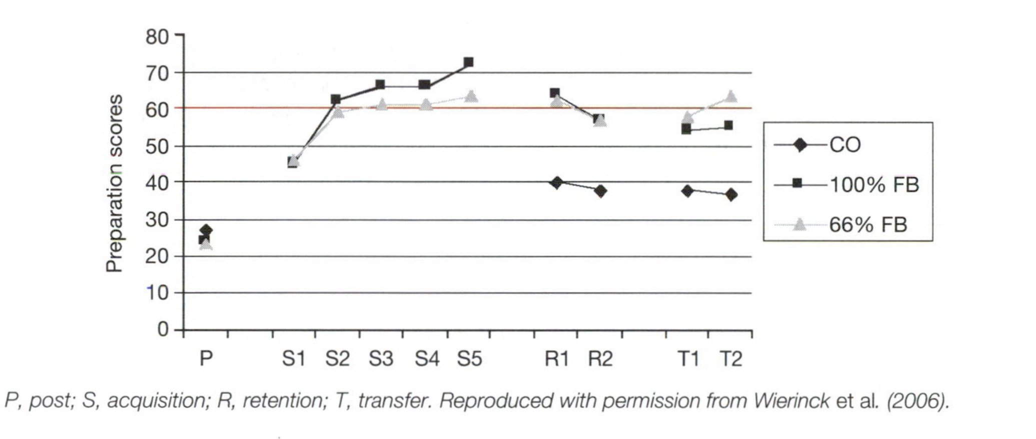

P, post; S, acquisition; R, retention; T, transfer. Reproduced with permission from Wierinck et al. *(2006).*

KP is information given to the performer that describes the quality of the movement pattern that led to the performance outcome. However, KP does not usually indicate anything about the product or goal achievement of the performance. The important point here is that KP differs from KR in terms of what aspect of performance the information refers to (Magill, 2001). If we take the hockey situation alluded to above, if giving KP the teacher would tell the student about the angle of the head of the hockey stick as it struck the ball. Thus, the teacher would be verbally augmenting the task's intrinsic information by telling the student what he or she did in order for the ball to miss the target (i.e. the goal).

Different forms of KR and KP

KR and KP can be given in many ways, and the form of feedback used is dependent on many factors such as the skill level of the person or the complexity of the task being completed. Verbal feedback is commonly used in a clinical or classroom setting. Consider the clinician who is helping a patient learning to locomote with an artificial limb, or the teacher instructing a class of 11-year-olds how to shoot a basket. In these situations verbal feedback is probably most appropriate. If you work with expert athletes you may wish to use videotape feedback, as is commonly used by the trainers of boxers who will replay their last fight to show them how they did. Many clinicians use biofeedback to provide information about the physiological process involved in performing skill, while more recent work has shown the potential usefulness of this tool as a training aid for high-class athletes. The discussion that follows focuses on the variety of alternative techniques available to the coach or clinician, and when best to use them depending upon the organismic, task or environmental constraints.

Research Highlight: Using feedback in a clinical setting

Augmented feedback, when properly employed, may have practical implications for rehabilitation therapy since the re-acquisition of motor skills is an important part of functional recovery (Winstein, 1991). Feedback is particularly important for those patients who may have cognitive, perceptual and neurological impairments who are unable to use intrinsic feedback. Cirstea *et al.* (2006) examined reaching in patients with chronic hemiparesis due to stroke. The patients were randomly assigned to three groups. Group 1, KR, with 20% concerned with precision, and group 2, KP about the joint motions, were asked to practice a reaching task which involved 75 repetitions per day, 5 days a week for 2 weeks. Group 3 practiced a non-reaching task. Groups 1 and 2 showed greater gains in terms of their motor function than the control group. Equally important was the finding that those patients who were the more severely impaired and had received KP, showed greater improvements in terms of their memory and planning abilities.

Verbal feedback

Verbal KP should be based on the most critical error made during a practice attempt; the professional should identify this based on a skill analysis and a prioritized list of components (Thorpe and Valvano, 2002). Verbal feedback can be an effective tool and even incorrect verbal feedback has been shown to override visual feedback in some situations (van Vliet and Wulf, 2006). Verbal feedback can be either or both descriptive and prescriptive. Descriptive verbal feedback tells the performer what kind of error they have made, for example 'You let go of the ball too soon'. Prescriptive verbal feedback not only identifies the error but tells the person what to do to correct the error, 'You need to toss the ball above the height of your extended arm'. To date there is little empirical evidence that dictates which is better for motor skill learning, and in general the answer lies in the skill level of the performer. Most professionals agree that prescriptive feedback is generally better for novices, while descriptive feedback is more appropriate for expert performers. In general, descriptive KP statements are more useful once the performer knows what to do to make a correction and thus improve performance.

Videotape feedback

Visual feedback is commonly given by using videotape replays by the coach or therapist. Indeed, videotape replays are a common practice that many assume aids skill acquisition by providing the learner with a visual depiction of their action. The effectiveness of videotape feedback depends very much upon the skill level of the learner (*Fig. 3*). In general, it is less effective for novices unless critical information is pointed out and then supplemented with verbal cues

Fig. 3. Players can receive feedback from watching video recordings of their performance.

(Kernodle and Carlton, 1992). In addition, this kind of feedback is more effective if used over an extended period of time (Rothstein and Arnold, 1976).

Using the task of ball-throwing with the non-dominant limb, Janelle *et al.* (1997) examined whether video feedback was effective during acquisition of a novel task. Three groups of learners received movement-related information. One of the groups watched video of their technique after every five trials (SUMMARY). It was also suggested that learners would benefit most from choosing the schedule of video provision. Therefore, the other two groups consisted of a group that chose when feedback was given (SELF) and a group that had no choice but had a matched feedback schedule to the SELF group (YOKED). Each subject was filmed performing the acquisition trials and had access to outcome information as they saw where on the target the ball landed. The form scores from the Janelle *et al.* (1997) study lend clear support to the use of video feedback for learners (*Fig. 3*). A KR group, who received no video feedback and solely outcome-related information, consistently showed worse technique according to impartial judges than the three video groups (accuracy scores were also lowest in the KR group!). During skill-acquisition trials, the SUMMARY and SELF groups performed as well as each other; however, in retention trials without feedback, the better form scores came from the SELF group. Finally, the YOKED group appeared to have suffered from not being given the independence to choose when feedback was administered. Despite watching the same amount of video at the same times as the SELF group (only on 11% of trials), the retention of the YOKED group was not as good.

Kernodle and Carlton (1992) examined a throwing task using the non-dominant arm. During learning, feedback was presented in the form of knowledge of results, knowledge of performance, knowledge of performance with attention-focusing cues, or knowledge of performance with error-correcting transitional information via videotaped replays. Subjects practiced this skill over 600 trials and over a 4-week period and viewed a videotaped model after every 10th trial. Performance was assessed with respect to both throwing distance and throwing form. Kernodle and Carlton (1992) found that KR and KP alone were less beneficial than the videotape replays with additional information about the correction to be made and cues about how to focus on the important aspects of the movement. Movement form ratings followed the same trend. These results support the hypothesis that KR alone may not be the most potent form of feedback in multiple-degree-of-freedom activities, and that KR combined with additional information can lead to significant gains in skill acquisition.

Kinematic/kinetic displays

Traditional forms of KR can specify only what not to do on the next attempt. However, kinematic/kinetic feedback can provide information about what to do on the next attempt. Kinematic/kinetic feedback frequently involves displaying results graphically for the learner, usually limited to one or two parameters of performance, such as joint angular displacement. Due to its instrumentation, kinematic/kinetic feedback is generally used to facilitate skill improvement in experts. According to Gentile (2000), novices are still 'getting the idea of the movement' and may find the sophisticated information available to them from this type of feedback of little use and very confusing. As Newell and Walter (1981) suggest, videotape replay may provide too much information to the learner and it is important that providing cues to focus attention on critical aspects of the task are provided.

Although it is thought to be less effective when the skill to be learned requires multiple degrees of freedom, kinematic feedback has been found to be effective for learning a golf shot (Wood *et al.*, 1992), a bimanual coordination task (Hurley and Lee, 2006), and for improving the kinematic consistency of rowing (Anderson *et al.*, 2005). Anderson *et al.* (2005) used accelerometers to examine if real-time visual ipsative (comparison with oneself) feedback improves kinematic consistency in rowing. All rowers completed three 2000 m time trials on an ergometer with three different visual feedback interventions: no feedback, detailed feedback, and summary feedback. Anderson *et al.* (2005) deemed feedback of detailed kinematic information to enhance kinematic consistency significantly more than either no feedback or summary feedback interventions, although summary feedback was shown to enhance kinematic consistency more than no feedback.

Biofeedback: using electromyography and electroencephalography

Biofeedback is described as a technique that provides information about performance via a signal from an external source which is related to a physiological process (e.g. heart rate, brain activity, muscle activity). This enables performers to alter or improve their performance by using information provided by their own body; however, it must be presented in such a way that people do not become dependent upon it. Although the most common form of biofeedback is that which uses the activity of muscles, electromyography (EMG), it can comprise visual, auditory or tactile information. EMG feedback is commonly used in a clinical setting and has been shown to be effective in treating patients with a variety of motor difficulties; recent developments have meant that EMG biofeedback training devices can be used at home.

Petrofsky (2001) showed that EMG biofeedback was effective in reducing Trendelenburg gait, a disturbance of walking pattern which, during the stance phase of walking, causes the pelvis to tilt down on the opposite side and the limb struggles to clear the ground. It is very important to strengthen the muscles around the hip and in the leg (to prevent pathologies due to altered gait). However, as a therapist spends little time with a patient they often walk incorrectly when the therapist is not present. Petrofsky (2001) studied two groups of patients with Trendelenburg gait due to incomplete spinal cord injury, receiving therapy (muscle strengthening and gait training) for 2 hours a day, 5 days a week in a clinic. Biofeedback was also accomplished for 30 minutes of the training on each patient using EMG. In addition, five patients wore an EMG biofeedback training device at home and thus received continuous biofeedback therapy every time they walked, and not biofeedback limited to only 30 minutes a day. The device provided warning tones giving feedback of improper gait through bilateral assessment of the use of the gluteus medius muscles. Patients undergoing clinical therapy showed about a 50% reduction in hip drop due to therapy. However, the group that used the home training device showed almost normal gait after the 2-month period.

Many researchers have begun to examine the use of electroencephalography (EEG) as a form of biofeedback. EEG is the neurophysiological measurement of the electrical activity of the brain by recording from electrodes placed on the scalp. The resulting traces are known as an electroencephalogram (EEG) and represent an electrical signal (postsynaptic potentials) from a large number of neurons.

Previous research has examined the EEG of expert sportsmen and found that they exhibit a distinct pattern of brain activity from that seen in novices (Radlo *et al.*, 2002). By examining EEG state prior to and during sporting events in experts, it allows researchers to examine the use of neurofeedback (biofeedback using EEG) to mimic cortical patterns in non-experts in an attempt to enhance sport performance (Vernon, 2005). For example, researches have found an increase in EEG activity in the left side of the brain prior to performance of the skill in experts. This is thought to represent a reduction in cortical activation in the left hemisphere of the brain, reducing the verbalizations of the behavior and allowing the visuospatial processes of the right hemisphere of the brain to become more dominant (Salazar *et al.*, 1990). These findings led Landers *et al.* (1991) to find out whether using neurofeedback novices could reduce the level of activity of the left hemisphere of the brain and if that affected performance. Three groups of pre-elite archers received either no neurofeedback (control) or neurofeedback that was to be used to reduce the activity of either the left or right hemisphere of the brain. Pre- and post-training, all archers completed a total of 27 shots from a distance of 45 m, with the level of performance measured as the distance between the arrow and center of the target They found that those that trained to shift the level of cortical activity towards more negativity in the left hemisphere of the brain showed a significant improvement in performance, whereas those that had trained to shift the level of cortical activity towards more negativity in the right hemisphere showed a significantly poorer performance; no difference was noted in the control group. However, examination of the participants' EEG data failed to reveal a clear pattern of change due to neurofeedback, leading Landers *et al.* (1991) to conclude that their results provide some support for neurofeedback as a method of enhancing the performance of pre-elite archers.

Important considerations for giving augmented feedback

Sport and exercise scientists and therapists alike agree that several timing characteristics of augmented feedback can affect skill learning. The first important point to consider is whether the learner should receive the augmented feedback during or after a skilled performance. Take, for example, the child learning to kick a ball. The instructor could give feedback during the kick, after the ball has been kicked, or both. Secondly, one needs to consider how the length of an activity during time intervals preceding and following the presentation of augmented feedback affects skill learning. If we go back to our novice footballer, the first time interval is that from the child kicking the ball and being given augmented feedback, and the second time interval is the time that elapses between you giving augmented feedback and them attempting to kick the ball again. Finally, for the instructor, the frequency with which we give augmented feedback needs to be considered – our novice footballer could receive feedback after each strike of the ball or each trial, or after, say, three or five trials. The points raised above leave us with three important questions that need to be answered:

1. does the timing of feedback really matter?
2. does it matter how long a person must wait before they get feedback or how long before having another trial they get feedback?
3. is the frequency with which we give augmented feedback an important consideration for the instructor or therapist?

The answer to each of these questions is probably an emphatic yes, and the next part of this section will help you understand why.

Box 1. Major characteristics of concurrent and terminal feedback

Concurrent feedback

- Feedback that is given during the movement
- Feedback provided in real time
- Possible direct involvement of the performer with the training device
- Better for complex routines
- Could be used during a closed-loop movement
- May compromise attention allocation availability in novices.

Terminal feedback

- Feedback that is given after the movement
- Could be used during an open-loop movement
- Can be better for beginners and clinical populations
- Places less demand on attention.

The timing of feedback

The first of the issues that we must consider with respect to giving augmented feedback is whether we give it while the person is performing a movement, which is referred to as concurrent augmented feedback, or to give it at the end of the practice attempt, which we call terminal augmented feedback. To date the question of whether terminal or concurrent feedback is better for facilitating skill acquisition has not been fully answered. If we remember that one of the important facts about augmented feedback is that it supplements the intrinsic feedback we receive when performing a movement, it seems reasonable to assume that if we are unable to interpret intrinsic feedback, then concurrent augmented feedback will probably be best. Several studies have revealed that the performance and learning of complex tasks, such as dance routines (Clarkson *et al.*, 1986), swimming (Chollet *et al.*, 1988) and gymnastics moves (Baudry *et al.*, 2006), is enhanced when concurrent augmented feedback is available.

The KR-delay and post-KR intervals

Another important issue related to the timing of augmented feedback concerns when the feedback is given terminally. The KR-delay interval is the time between the completion of a trial and the instructor giving KR. The post-KR interval is the time between the coach giving KR and the performer attempting the next trial (*Fig. 4*). A common assumption is that the shorter the KR-delay interval the better for the learning. However, if learners receive this information too soon after completion of a movement, they are not able to engage in the subjective analysis of task-intrinsic feedback. Subsequently, performers' ability to develop their error-detection capability is compromised. Delaying the presentation of augmented feedback does not appear to affect skill learning; however there does seem to be a minimum amount of time that must pass before it is given (Swinnen *et al.*, 1990).

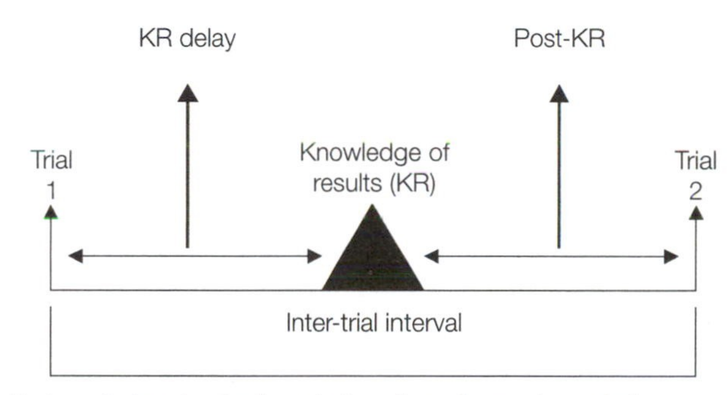

Fig. 4. The timing of when to give knowledge of results can be varied.

The KR-delay interval is not the only consideration for the coach or clinician when deciding when to give terminal feedback. The second factor that can affect skill learning is the type of activity the performer engages in during the KR-delay interval, and overall researchers have found three types of outcome associated with different types of activity.

In general, skill learning is not affected when activity is undertaken during the KR-delay interval. There is some evidence, although it is somewhat limited, that suggests that activity during the KR-delay interval can hinder skill learning. Most importantly, research that has shown this has also provided an insight into what types of situations this outcome occurs in. Using an experimental design in which subjects had to learn either a cognitive or a motor task during the KR-delay interval, Martenuik (1986) showed that if the task undertaken interfered with the exact learning processes required by the primary task, then motor skill learning was hindered. The other type of KR-delay interval activity that has showed a similar effect of learning is one that involves estimating the movement-time error of another person's lever movement, which the second person performed during the interval (Swinnen *et al.*, 1990).

Interestingly, this same work showed that if the performer estimates his or her own error, in what is called subjective error estimation, before receiving augmented feedback, it benefited motor skill learning. Liu and Wrisberg (1997) reported similar results when participants had to learn a throwing task. Participants who estimated their movement form during acquisition produced significantly higher throwing accuracy and lower estimation error during retention than those who did not. This attenuation in motor skill leaning is probably due to learners testing hypotheses about their performance based upon their own error estimation, with the KR given to them by their coach (Guadagnoli and Kohl, 2001)

Hurley and Lee (2006) looked at the influence of augmented feedback on the acquisition of a bimanual coordination pattern. They found that learning was greatly affected by augmented feedback and that ongoing augmented feedback was more effective than terminal feedback in terms of skill transfer (*Fig. 5*). However, it should be noted that in other studies ongoing feedback has been found to be detrimental to skill transfer (Salmoni *et al.*, 1984; Schmidt and Wulf, 1997). Of equal importance to the skill-acquisition process is the length of the post-KR interval. Remember, this is the time period between having received KR and performing the next trial. It is therefore assumed that this period of time is of importance as it allows the performer to develop a plan of action based upon both the augmented KR and the sensory feedback inherent to performing the task. Prior to the formation of this plan of action, the performer has to process the feedback available to them after completing the trial. It thus seems logical to assume that performers need a minimum length of time to do this and

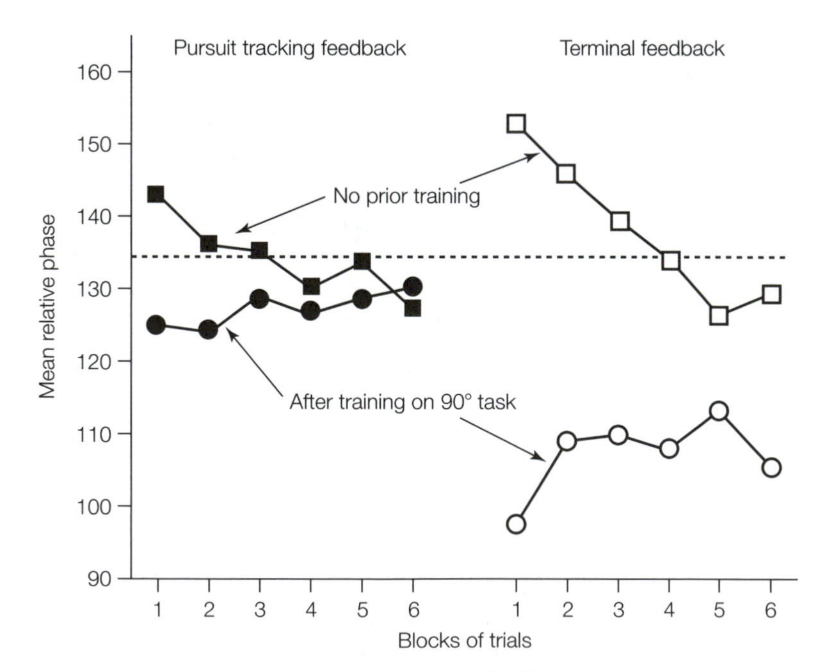

Fig. 5. Performance of the four groups. Reproduced with permission from Hurley and Lee (2006), p.345, Figure 2.

thus need a minimum length of time for the post-KR interval. Although there is no evidence indicating an optimum length for the post-KR interval, a lot of empirical work shows that in general the post-KR interval is far too short and that the shorter the post-KR interval the poorer the learning.

As research has shown that the type of activity engaged in during the KR-delay interval can benefit, interfere with or have no effect on learning, empirical work has also shown the type of activity undertaken during the post-KR interval produces similar effects. Very little work has demonstrated the beneficial learning effects that can result from undertaking activity during the post-KR interval (Magill, 1988). However, several researchers have reported results indicating that activity during the post-KR interval either hinders learning or has no effect on skill learning; unfortunately, much of the work had serious methodological flaws that do not permit us to draw accurate conclusions.

The data that have been reported do allow us to make some assumptions about the learning process. In general, they support the view we discussed earlier that learners engage in important planning activities during the post-KR interval. More specifically, during the post-KR interval, learners compare the augmented KR with the intrinsic feedback associated with completing the task and use this to determine how to execute the next attempt at performing the skill. However, this does not explain the beneficial effect on learning that attempting to learn another motor skill during the post-KR interval has. Magill (2001) hypothesizes that these post-KR interval motor learning tasks increase the problem-solving activity the performer has to undertake, which enables the performer to transfer more successfully to a situation that requires similar problem-solving activity.

Precision of augmented feedback

Another factor that must be considered when deciding what type of augmented feedback to provide to the learner is the level of precision associated with the feedback. When we talk about precision we are referring to the degree of exactness of the information we give to the learner, and this information is in either a qualitative or quantitative format. Qualitative feedback generally informs the performer if their performance was correct or incorrect and is provided by information that tells them if their performance was too fast/slow, too hard/soft, too long/short. Quantitative feedback indicates the direction and magnitude of the error and is presented with varying degrees of precision.

As you may expect, as the precision of feedback increases, so too does the amount of processing required of the performer, and logically the time required for this processing also increases. Research has consistently shown that increasing the level of precision associated with feedback is advantageous to the performer up until a certain point, beyond which performance is negatively affected (*Box 2*).

Box 2. Key points concerning precision of feedback (Newell, 1991)

- As the precision of feedback increases so too should the time provided to process the information.
- Novice performers should receive less precise feedback than intermediate performers.
- More-general infcrmation about the learner's performance is more desirable in the early learning stages.

Frequency of augmented feedback

How often should an instructor or clinician provide augmented feedback? In line with Adams' theory (Section H), initial studies done over 30 years ago showed that the more KR a performer received the better the performance (Adams, 1971; Adams *et al.*, 1972). In one study (Bilodeau and Bilodeau, 1958a), participants were attempting to learn a lever-pulling task and were provided with KR after every third or fourth trial, or after every trial. At the end of the practice, the group that had received KR after every trial exhibited the least amount of performance error. However, in this study the two groups received different amounts of practice in that the group receiving KR after every trial practiced the task only 10 times, while the other groups had 40 practice trials.

Later work started to suggest that the conclusion that 'the more KR the better' was not always accurate. As Rose and Christina (2005) note, two aspects of the earlier studies (Bilodeau and Bilodeau 1958a, b) were altered during these later studies and these could account for the reversal of findings. In the later work, no-KR retention tests were included, which made it possible to differentiate performance from learning effects. Secondly, absolute frequency of KR (the total number of trials after which KR was provided) was no longer held constant but was allowed to vary with the relative frequency of KR (the proportion of trials after which KR was provided). This second change eliminated the problem with the earlier work of Bilodeau and Bilodeau (1958b) in which the participants received different amounts of practice.

Fading-frequency schedules of KR

For many years there was a long-held belief by many experts in motor learning that the absolute frequency of augmented feedback (i.e. the number of times that feedback was provided) was a critical variable in learning, but that relative frequency (the ratio of feedback-provided trials to the total number of trials) was not an important variable for learning. However, in their review of the motor learning literature, Salmoni *et al.* (1984) found that those studies that had included retention tests following relative frequency manipulations showed that relative frequency might indeed be an important variable in learning. The general trend of the data showed that low relative frequencies might be better than high relative frequencies of augmented feedback.

An overall relative frequency (say 50%) has higher relative frequency (say 75%) in early practice and a lower relative frequency (say 25%) later in practice. This 'fading' technique has been found to be more effective for learning than a 100% relative frequency throughout practice.

Winstein and Schmidt (1990) conducted three experiments that examined how reducing the frequency of knowledge of results enhances motor skill learning. In experiment 1 subjects received 100% or 50% feedback, in experiments 2 and 3 the group that received 50% feedback had the amount of feedback gradually reduced. It was shown that a reduced KR was as effective on a retention test as full KR, in experiment 1. In experiments 2 and 3 it was found that when the scheduling of KR was manipulated so that the frequency of KR trials was gradually reduced, learning was enhanced. *Figure 6* shows the results from experiment 2.

An alternative method of providing knowledge of results is by a method known as bandwidth knowledge of results. By this method the nature of augmented feedback is determined by a bandwidth or acceptable range of error. The first reported experiment to use this method was carried out by Sherwood

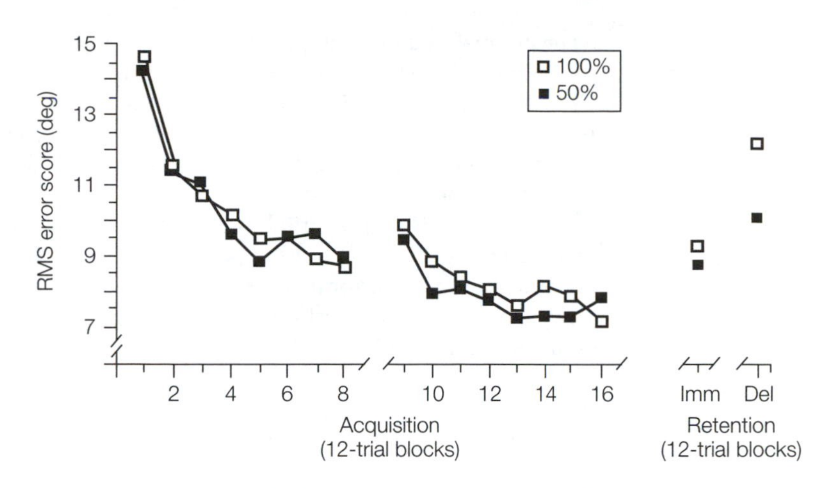

Fig. 6. Average root-mean (RMS) for two groups, one receiving 100% KR the other 50% KR. Imm, immediate feedback; Del, delayed feedback. Reproduced with permission from Winstein and Schmidt (1990), p.683, Figure 3.

(1988). Using a movement timing task, Sherwood provided subjects with augmented feedback only when their errors exceeded certain tolerance limits about the goal. For one group of subjects, augmented feedback was provided on every trial (a '0% bandwidth'). For other subjects, augmented feedback was provided only when performance error was greater than ±5%, or greater than ±10% of the goal. By this manipulation, Sherwood also influenced the frequency schedule by which augmented feedback was provided (since feedback was not provided for trials on which performance was within the bandwidth limits). When using this technique, augmented feedback is only provided if errors in performance exceed a certain range. The KR serves both an informational and motivating/reinforcing function. The learner may well interpret absence of KR after a practice attempt as satisfactory performance.

Early in practice, when many errors occurred, the relative frequency of augmented feedback was fairly high. However, as performance improvements were made over practice, the relative frequency became less frequent. Performance on a retention test was best for the group that had received the 10% bandwidth and poorest for the 0% bandwidth group, with the 5% group performing at an intermediate level. At a minimum, these findings are consistent with the effects of reduced relative frequency, although it seems that the bandwidth manipulation has an additional learning effect. The other similarity is that 'fading' the relative frequency of feedback over practice (giving frequent feedback early in practice and infrequently later) is both an effective strategy and one that occurs as a natural consequence of the bandwidth procedure.

Summary feedback

Summary feedback involves withholding feedback for a given number of practice attempts, with the optimal summary KR length being influenced by the complexity of the task to be learned and the skill level of the learner. Schmidt *et al.* (1989) illustrate the effects of summary feedback on learning and performance. They divided participants into four groups according to the number of trials that were summarized on a graph and presented to the subject as augmented feedback. In the 'Sum 1' group, feedback was provided after every trial. In the 'Sum 5', 'Sum 10' and 'Sum 15' groups, feedback was provided on a graph about the 5, 10 or 15 preceding trials, illustrating how performance either changed or remained consistent over trials. Immediate and delayed tests of retention followed the practice period.

Schmidt *et al.* (1989) reported two important findings. Firstly, the data showed that the larger the summary the poorer the acquisition performance. Secondly, delayed retention performance was achieved by the 'Sum 15' and 'Sum 10' groups, and the worst performance by the 'Sum 5' and 'Sum 1' groups. These findings are particularly interesting because of the opposite effects that the summary conditions had on the practice performance and on learning (Schmidt *et al.*, 1990)

Schmidt *et al.* (1990) conducted an experiment using an anticipatory timing task that aimed to examine the influence of summary length on skill learning. Knowledge of performance was provided after 1, 5, 10, or 15 trials. The results of this experiment, seen in *Fig. 7*, showed that an increase in summary length negatively influenced the performance in acquisition. However, when examining the transfer tests it can be seen that a delay of five trials was the most effective for learning.

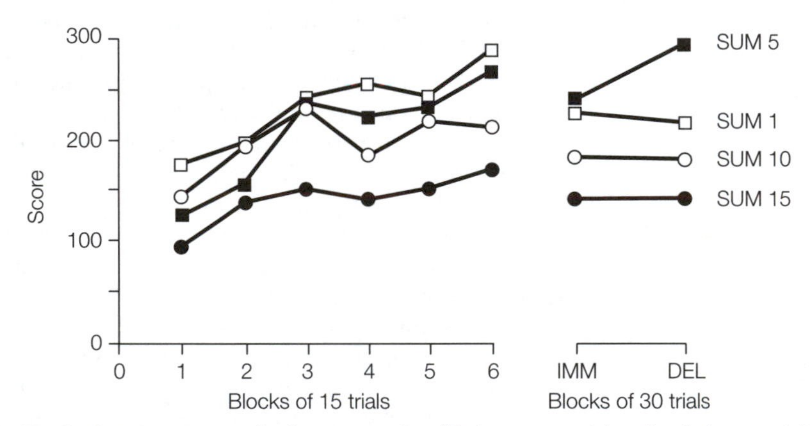

Fig. 7. Performance scores for four groups for differing summary lengths during acquisition and transfer tests on an anticipatory timing task. Reproduced with permission from Schmidt et al. *(1990), p.334, Figure 3).*

Because augmented feedback is such an important part of motor skill learning, it is important to understand what information the therapist or instructor should provide to facilitate learning and how often he or she should give the information. When giving feedback there are three important things to consider. Firstly, he or she should decide the precision of the information. Augmented feedback can either be too precise or too general to aid learning. Secondly, he or she should determine the content of the feedback. Augmented feedback directs attention to parts of the skill; therefore, augmented feedback should direct attention to that part of the skill that is the most important to improve on the next trial. Thirdly, the professional should establish the form of presenting the augmented feedback. All of the above will also depend on the complexity of the skill, the age and the skill level of the learner.

Conclusion

This section has provided an overview of the role and function of feedback in motor learning and control. It should be remembered that as with many areas of motor control there is conflicting evidence as to the benefits of augmented feedback. The benefits tend to include enhanced goal attainment, and the detrimental effects can include dependence on feedback, which of course is not always available. It should also be remembered that feedback has a motivational role and that there are strong links between feedback and the discipline of psychology.

Further reading

Liu, J. and Wrisberg, C.A. (1997) The effect of knowledge of results delay and the subjective estimation of movement form on the acquisition and retention of a motor skill. *Res Q Exerc Sport* **68**, 145–151.

Russell, D.M. and Newell, K.M. (2006) On No-KR tests in motor learning, retention and transfer. *Hum Mov Sci* **26**, 155–173.

Wulf, G. and Shea, C.H. (2004) Understanding the role of augmented feedback: the good, the bad, and the ugly. In: Williams, A.M. and Hodges, N.J. (eds) *Skill Acquisition in Sport: research, theory and practice.* Routledge, London, pp. 121–144.

References

Adams, J.A. (1971) A closed-loop theory of motor learning. *J Mot Behav* **3**, 111–149.

Adams, J.A., Goetz, E.T. and Marshall, P.H. (1972) Response feedback and motor learning. *J Exp Psychol* **92**, 391–397.

Anderson, R., Harrison, A. and Lyons, G.M. (2005) Accelerometry-based feedback – can it improve movement consistency and performance in rowing? *Sport Biomech* **4**, 179–195.

Baudry, L., Leroy, D., Thouvarecq, R. and Coller, D. (2006). Auditory concurrent feedback benefits on the circle performed in gymnastics. *J Sport Sci* **24**, 149–156.

Bilodeau, E.A. and Bilodeau, I.M. (1958a) Variable frequency of knowledge of results and the learning of a simple skill. *J Exp Psychol* **55**, 379–383.

Bilodeau, E.A. and Bilodeau, I.M. (1958b) Variations in temporal intervals among critical events in five studies of knowledge of results. *J Exp Psychol* **55**, 603–612.

Chollet, D., Micallef, J. and Rabischong, P. (1988) Biomechanical signals for external biofeedback to improve swimming techniques. In: Ungerechts, B.E., Wilke, K. and Reischle, K. (eds) *Swimming Science V*. Human Kinetics, Champaign, IL, pp. 389–396.

Cirstea, C.M., Ptito, A. and Levin, M.F. (2006) Feedback and cognition in arm motor skill reaquisition after stroke. *Stroke* **37**, 1237.

Clarkson, P.M., James, R., Watkins, A. and Foley, P. (1986) The effect of augmented feedback on foot pronation during bar exercise during dance. *Res Q Exerc Sport* **57**, 33–40.

Gentile, A.M. (2000) Skill acquisition: action, movement, and neuromotor processes. In: Carr, J.H. and Shephard, R.B. (eds) *Movement Science: foundations for physical therapy*, 2nd Edn. Aspen, Rockville, MD, pp.11–187.

Guadagnoli, M.A. and Kohl, R.M. (2001) Knowledge of results for motor learning: relationship between error estimation and knowledge of results frequency. *J Mot Behav* **33**, 217–224.

Hartveld, A. and Hegarty, J. (1996) Augmented feedback and physiotherapy practice. *Physiotherapy* **2**, 480–490.

Hurley, S.R. and Lee, T.J. (2006) The influence of augmented feedback and prior learning on the acquisition of a new bimanual coordination pattern. *Hum Mov Sci* **25**, 339–348.

Janelle, C.M., Barba, D.A., Frehlich, S.G., Tennant, K. and Cauraugh, J.H. (1997) Maximizing performance feedback effectiveness through videotape replay and a self-controlled learning environment. *Res Q Exerc Sport* **68**, 269–279.

Kernodle M.W. and Carlton L.G. (1992) Information feedback and the learning of multiple-degree-of-freedom activities. *J Mot Behav* **24**, 87–196.

Landers, D.M., Petruzzello, S.J., Salazar, W. *et al.* (1991) The influence of electrocortical biofeedback on performance in pre-elite archers. *Med Sci Sports Exerc* **23**, 123–129.

Liu, J. and Wrisberg, C.A. (1997) The effect of knowledge of results delay and the subjective estimation of movement form on the acquisition and retention of a motor skill. *Res Q Exerc Sport* **68**, 145–151.

Magill, R.A. (1988). Activity during the post-knowledge of results interval can benefit motor skill learning. In: Meijer, O.G. and Roth, K. (eds) *Complex Movement Behavior: the motor action controversy*. Elsevier Science Publishers, Amsterdam, pp. 231–246.

Magill, R.A. (2001) *Motor Learning*, 6th Edn. McGraw-Hill Higher Education, Boston.

Martenuik, R.G. (1986) Information processes in movement learning: capacity and structural interference effects. *J Mot Behav* **18**, 55–75.

Newell, K.M. (1974) Knowledge of results and motor learning. *J Mot Behav* **6**, 235–244.

Newell, K.M. (1991) Motor skill acquisition. *Annu Rev Psychol* **42**, 213–237.

Newell, K.M. and Walter, C.B. (1981) Kinematic and kinetic parameters as information feedback in motor skill acquisition. *J Hum Mov Stud* **7**, 235–254.

Petrofsky, J.S. (2001) The use of electromyogram biofeedback to reduce Trendelenburg gait. *Eur J Appl Psychol* **85**, 491–495.

Radlo, S.J., Steinberg, G.M., Singer, R.M., Barba, D.A. and Melinkov, A. (2002) The influence of an attentional focus strategy on alpha brain wave activity, heart rate, and dart throwing performance. *Int J Sport Psychol* **33**, 205–217.

Rose, D.J. and Christina, R.W. (2005) *A Multilevel Approach to the Study of Motor Control and Learning*, 2nd Edn. Pearson, San Francisco.

Rothstein, A.L. and Arnold, R.K. (1976) Bridging the gap: application of research on videotape feedback and bowling. *Motor Skills: Theory Into Practice* **1**: 36–63.

Salazar, W., Landers, D.M., Petruzello, S.J., Crews, D.J. and Kubitz, K.A. (1990) Hemisphere asymmetry, cardiac response, and performance in elite archers. *Res Q Exerc Sport* **61**, 251–359.

Salmoni, A.W., Schmidt, R.A. and Walter, C.B. (1984) Knowledge of results and motor learning: a review and critical reappraisal. *Psychol Bull* **95**, 355–386.

Schmidt R.A., Lang, C.A. and Young, D.E. (1990) Optimizing summary knowledge of results for skill learning. *Hum Mov Sci* **9**, 325–348.

Schmidt, R.A. and Wulf, G. (1997) Continuous concurrent feedback degrades skill learning: implications for training and simulation. *Hum Factors* **39**, 509–525.

Schmidt, R.A., Young, D.E., Swinnen, S. and Shapiro, D.C. (1989) Summary knowledge of results for skill acquisition: support for the guidance hypothesis. *J Exp Psychol Learn Mem Cogn* **15**, 352–359.

Sherwood, D.E. (1988) Effect of bandwidth knowledge of results on movement consistency. *Percept Mot Skills* **66**, 535–542.

Shumway-Cook, A., Anson, D. and Haller, S. (1988) Postural sway biofeedback for pretatraining postural control following hemiplegia. *Arch Phys Med Rehabil* **69**, 395–400.

Swinnen, S.P., Schmidt, R.A., Nicholson, D.E. and Shapiro, D.C. (1990) Information feedback for skill acquisition: instantaneous knowledge of results degrades learning. *J Exp Psychol Learn Mem Cogn* **16**, 706–716.

Thorpe, D.E. and Valvano, J. (2002) The effects of knowledge of performance and cognitive strategies on motor skill learning in children with cerebral palsy. *Paed Phys Ther* **14**, 2–15.

Van Vliet, P.M. and Wulf, G. (2006) Extrinsic feedback for motor learning after stroke: what is the evidence? *Disabil Rehabil* **28**, 831–840.

Vernon, D.J. (2005) Can neurofeedback training enhance performance? An evaluation of the evidence with implications for future research. *Appl Psychophysiol Biofeedback* **30**, 347–364.

Wierinck, E., Puttemans, V. and van Steenberghe, D. (2006) Effect of tutorial input in addition to augmented feedback on manual dexterity training and its retention. *Eur J Dent Educ* **10**, 24–31.

Winstein, C.J. (1991). Knowledge of results and motor learning – implications for physical therapy. *Phys Ther* **71**, 140–149.

Winstein, C.J. and Schmidt, R.A. (1990) Reduced frequency of knowledge of results enhances motor skill learning. *J Exp Psychol Hum Learn Mem Cogn* **16**, 677–691.

Wood, C.A., Gallager, J.D., Martino, P.V. and Ross, M. (1992) Alternate forms of knowledge of results: interaction of augmented feedback modality on learning. *J Hum Mov Stud* **22**, 213–230.

Section M

MOTOR DEVELOPMENT

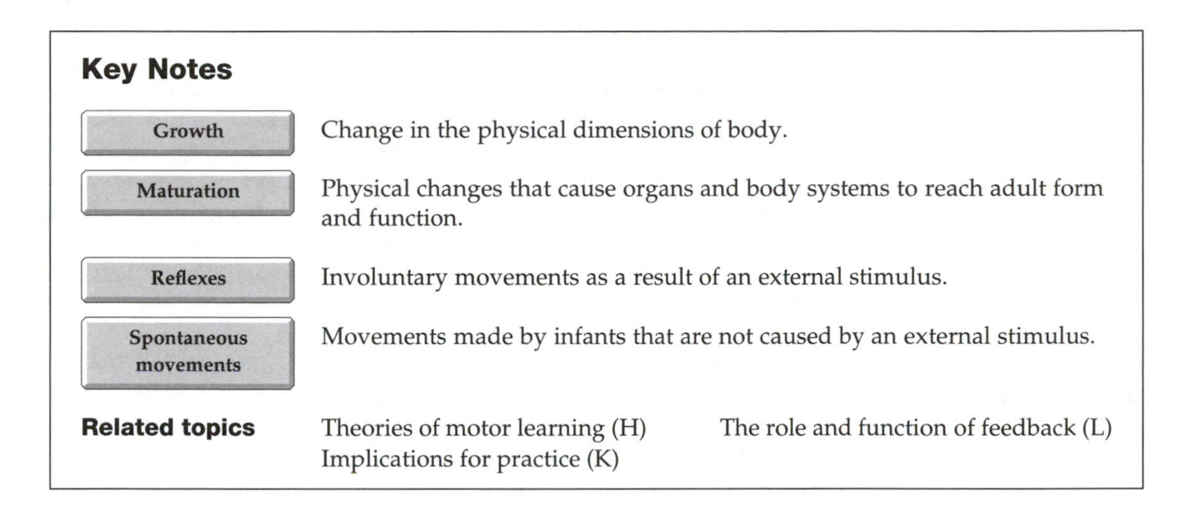

Key Notes

Growth	Change in the physical dimensions of body.
Maturation	Physical changes that cause organs and body systems to reach adult form and function.
Reflexes	Involuntary movements as a result of an external stimulus.
Spontaneous movements	Movements made by infants that are not caused by an external stimulus.

Related topics

Theories of motor learning (H)
Implications for practice (K)
The role and function of feedback (L)

Keogh and Sugden (1985) define development as an adaptive change towards competence and this definition is not dissimilar to that of Gallahue (1982), who states that development is 'A progressive change in motor control and motor behavior brought about by the interaction of both maturation and experience' (p.19). The understanding of motor development is based upon the integration of many behavior changes that take place within and across phases of development. As we shall see in Section N, a number of researchers have defined the phases of development for a variety of skills. The development of a child's motor behavior is a sequential process. This development starts with primitive reflexes then progresses to the learning of postural movements, next to locomotor responses, and finally to a range of manipulative movements.

The scientific study of motor development can be traced back to the 1930s and 1940s when developmental scientists such as Shirley (1931), Gessell (1939) and McGraw (1932, 1940, 1945) spent many years observing and reporting how infants gain control of their movements; these so called maturation models were popular for a long time. During the 1970s and 1980s, models of cognitive theorists were popular. This information-processing approach described motor control in terms of separate processes that can be categorized as input, central, output and feedback processes involving cognitive concepts, such as attention and memory. This approach emphasized the hierarchical nature of motor development, and attributed motor coordination to the establishment of schema, representations or motor programs in the higher centers that are responsible for the control of movement (Schmidt, 1975). Although this helped popularize motor behavior as a valid field of study, these models did not provide totally satisfactory accounts for motor development (Sugden and Wright, 1998). This hierarchical approach has lost popularity, and more recently the focus of attention has moved away from these models towards explanations that are more dynamical and ecological in nature. A dynamical systems approach examines the interaction between the demands made by internal constraints, such as body mechanics or the task, and external constraints such as the environment.

Dynamical systems theory offers a set of principles for studying emergence of behavior and a theory that better explains change (Turvey and Fitzpatrick, 1993; Thelen, 1995).

The study of motor development has had renewed interest during the last two decades. This work not only describes children's behavior in further detail and with greater accuracy, but more importantly using new ideas and exciting theoretical concepts to explain how and why changes take place that to date have been insufficiently explored. Keogh and Sugden (1985) note two basic questions in developmental psychology. The first question concerns describing change and how children differ at different phases in their lives; the second question is concerned with identifying the agents of these changes. This section considers the different theories that have been proposed to answer these questions. In addition, it will also consider two important aspects of motor development: reflexes and spontaneous movements.

Maturational perspective

Some of the most methodologically elegant and theoretically generative early work in the scientific study of motor development comes from the motor domain and researchers such as Shirley (1931), McGraw (1932, 1940, 1945) and Gesell (1939). These early investigators had varied motives for studying motor development: to establish developmental norms, to resolve the nature–nurture debate and to understand the developmental process and its underlying neurological basis. They began by describing the progression of behavioral norms leading to clear developmental milestones. Thus the maturational perspective emerged with descriptive studies of how the bodies of infants grow and change. Arnold Gesell's efforts led to catalogs of motor types, he described 23 stages and 28 subtypes of prone behavior producing similar catalogs for nearly all motor behaviors (Gesell and Ames, 1940). McGraw (1945) complemented this approach to motor development and described 'seven stages of erect locomotion'. She proposed that infants progressed sequentially through identifiable stages of locomotor behavior, the first being a simple stepping reflex, to the last phase of fully integrated walking (*Fig. 1*).

Both McGraw and Gesell considered motor processes as the primary driver of developmental change. For example, McGraw suggested that motor changes were driven primarily by changes in the motor cortex and its inhibitory influences on lower brain centers. Therefore the first assumption of this neural maturational approach is that the cerebral cortex is the agent of diverse, plastic and purposeful behavior. She stated that newborns are dominated by 'primitive subcortical nuclei' (McGraw, 1945, p.10), and that their behavior is therefore largely reflexive. Reflexes therefore formed an important part of these early theories. Before considering theories of development in detail we will start by looking in depth at infant reflexes. During this section we will consider the role of reflexes from a number of perspectives.

Reflexes

Newborn babies may appear to have very little control of their movements. However, at birth a child can produce a number of movement reflexes and indeed has some voluntary control (Von Hofsten, 1993). Kelso (1982) stated that to some extent at birth the final coordination of different muscles is already built into the system. A reflex is a movement response to a sensory stimulus such as touch, light or sound. Reflexes appear at a particular age, and most disappear (or are incorporated into movement) early in life, some are maintained throughout life. Infantile reflexes can be categorized into three types; primitive,

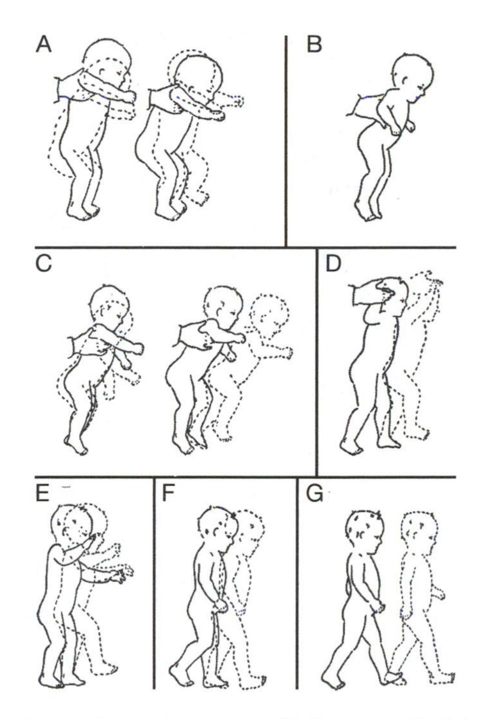

Fig. 1. The seven stages of erect locomotion (McGraw, 1945). Adapted from Thelen and Smith (1994).

postural, and locomotor (Haywood and Getchell, 2001). Early or primitive reflexes found in babies are related to infant survival and nourishment seeking as well as obvious protective mechanisms. These subcortically controlled reflex movements are present in all newborn babies (Peiper, 1963). The rate and strength at which they appear and disappear are considered indicators of healthy development or early indicators of central nervous system (CNS) disorders (Gallahue, 1982). There is some debate among researchers as to whether some reflexes disappear or are gradually incorporated into voluntary movements. For example, the role of reflexes in controlling early reaching and grasping is an area in which there has been much debate. This debate has focused on whether reaching and grasping develops as a result of the inhibition of primitive reflexes or through the integration of reflexes into voluntary movement. Hellebrandt *et al.* (1962) indicated that reflexes such as the tonic neck, tonic labyrinth and righting reflex underlie movements of everyday life in adults. This view is support by Twitchell (1970), who stated that reflexes remain and interact with voluntary controlled activity. Other researchers have argued that reflexes exist independently of voluntary control (McDonnel *et al.*, 1989). What is clear is that reflexes have a role in development. A summary of a number of primitive reflexes can be found in *Table 1*.

Primitive reflexes

There are many primitive reflexes found in infancy, with Prechtl (1977) identifying as many as 70. The function of primitive reflexes is one of survival; they therefore assist the infant in injury avoidance and feeding. The first reflexes to be considered are those involved in feeding. There are two main reflexes that are

Table 1. Primitive reflexes, their function, and age when present

Reflex	Function	Age when present
Rooting reflex: when the corner of the infant's mouth is stroked or touched, the infant will turn their head and open their mouth	Feeding	Up to 12 months
Sucking reflex: when the roof of the infant's mouth is touched, the baby will begin to suck	Feeding	Up to 6 months
Grasp reflex: contact with the infant's palm causes the infant to close their fingers in a tight grasp	Support	Up to 4 months
Withdrawal reflex: when an infant's limb makes contact with a sharp object the limb will be quickly withdrawn	Safety	Up to 12 months
Babinski reflex: when the sole of the foot (lateral side) is firmly stroked, the big toe bends towards the top of the foot, and the other toes fan out	Safety?	Up to 24 months
Moro reflex: the infant throws back their head and extends out the arms and legs	Safety	Up to 6 months
Startle reflex: the infant curls in and pulls in their arms and legs	Safety	Between 6 and 12 months
Tonic neck reflex: when the infant's head is turned to one side, the arm on that side stretches out and the opposite arm bends up at the elbow	Support	Up to 7 months

essential for feeding, the rooting reflex and the sucking reflex (*Fig. 2*). The rooting reflex enables the infant to locate the nipple by turning the infant's head on one side (this reflex also protects the child from suffocation). Contact near the baby's mouth will elicit the rooting reflex. Once the nipple is located, the sucking reflex is activated which enables the infant to get milk from the mother or feeding bottle. Other reflexes such as the grasp reflex and the withdrawal reflex are concerned with injury prevention. The grasp reflex can be seen when a baby takes tight hold of a finger or object placed in its palm (Twitchell, 1970). This enables the infant to keep a tight hold of its parent or any object that it wants or needs to hold onto. The withdrawal reflex can be seen when the infant comes into contact with an item that might cause it harm. For example, if the infant's foot contacts a sharp object the leg would be instantly pulled away. Other reflexes give an indication of the neurological health of the child. If these reflexes are lacking or are very weak it may indicate some form of dysfunction. The startle reflex, Moro reflex and tonic neck reflex for example, are used to assess dysfunction in conditions such as cerebral palsy. The startle reflex and Moro reflex occur when a sudden jolting movement or loud noise stimulates the infant. The startle reflex involves flexion of the limbs towards the trunk in a protective manner. The Moro reflex results in the limbs extending rapidly. This often occurs when an infant senses a loss in support and may well be an attempt to slow them down if they feel that they are falling. The tonic neck reflex occurs when an infant's head is turned to one side; the arm on that side stretches out and the opposite arm bends up at the elbow. This is often called the 'fencing' position and interestingly an example of this can be seen

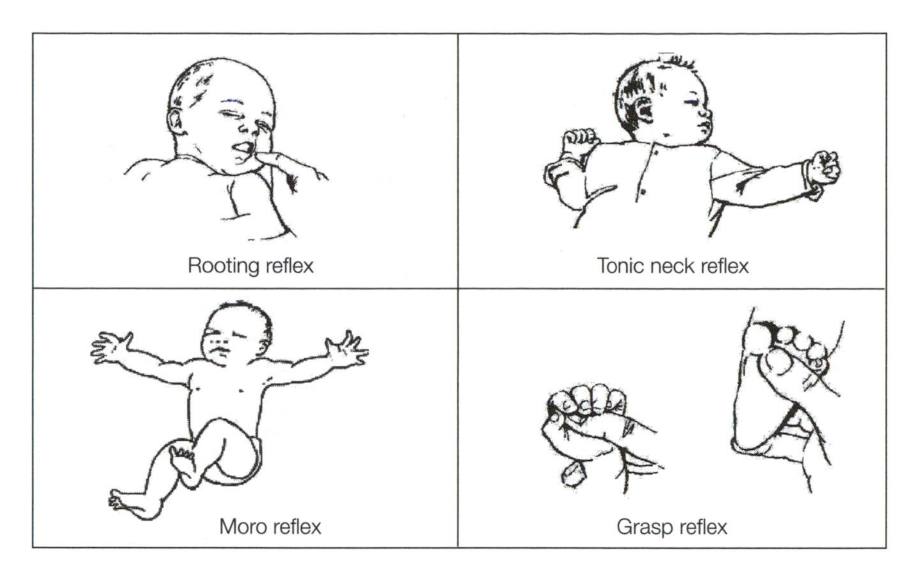

Fig. 2. Various primitive reflexes.

in adults where turning the head to one side results in limb extension on that side and flexion on the other, this can be seen when jumping to catch a ball with one hand. The final primitive reflex to be considered is the Babinski reflex. The Babinski reflex occurs when the big toe bends toward the top of the foot and the other toes fan out, after the sole of the foot has been firmly stroked. This is normal in younger children, but abnormal after two years of age.

Postural reflexes

Postural reactions or reflexes generally appear after the infant is two months old. By the end of the first year, or early in the second year, these reactions are no longer part of the infant's repertoire of movements. Postural reflexes can be divided into righting reflexes and equilibrium reflexes. Righting reflexes enable the infant to maintain a constant orientation in relation to the pull of gravity. Righting reflexes include the labyrinthine reflex, the neck-righting reflex, the pull-up reflex and the body-righting reflex. Equilibrium reflexes include the parachute reflex, the propping reflex and the Landau reflex.

The labyrinthine reflex can be elicited by holding the infant in an upright position and then tilting the infant forward. The infant should maintain the face in a vertical position and the mouth in a horizontal position. Another righting reflex is the neck-righting reflex which causes a range of responses. If the infant's head is extended (*Fig. 3*), then this will result in extension of the spine. Equally if the infant's is head is flexed the spine will flex. Finally, turning the infant's head to one side will result in reflex response in which the infant will turn his or her shoulders and body in the same direction as the head. The pull-up reflex is another righting reflex where the infant will pull themselves into an upright position if held by the arms. This keeps the head in an upright position. The next group of reflexes to be considered is equilibrium reflexes which involve a full body response when the infant experiences a change in the position of the centre of gravity.

The propping reflex is essential for the infant to be able to sit independently. This postural reflex develops at 5 to 7 months of age. Anterior propping actually

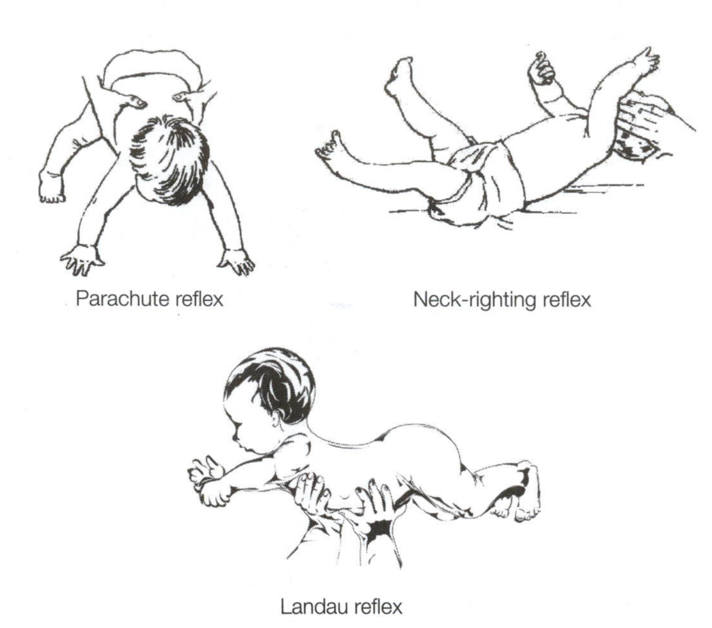

Fig. 3. Postural reflexes.

develops first, then lateral propping. For anterior propping the infant will extend the arms forward to prevent themselves falling forward, and lateral propping occurs when the infant is falling to one side or the other and extends the arm laterally to prevent this happening. The Landau reflex develops at around 4 to 5 months of age and can be elicited when the infant is suspended in the prone position (*Fig. 3*), the head will extend above the plane of the trunk. The trunk is straight and the legs are extended so the infant is opposing gravity. When the head is gently flexed, the legs drop into flexion. When the head is released, the head and legs will return to the extended position. Finally, the parachute reflex is the last of the postural reflexes to develop. It usually appears at 8 to 9 months of age. When the infant is turned face down towards the floor, the arms will extend as if the infant is trying to catch themselves. All of these reflexes enable the infant to maintain and correct changes to the body's posture.

Locomotor reflexes

Finally there are three locomotor reflexes: stepping, swimming and crawling; these appear much earlier and then disappear (or are incorporated into movement) before the onset of voluntary behavior. If a child is held so that their feet can make contact with a surface, they will make a stepping action that resembles walking; this stepping reflex can be seen in the first few months and it disappears at around 5 to 6 months of age. A child will also crawl if placed on the floor in a prone position, and the crawl reflex can be seen for the first 3 to 4 months. The final locomotor reflex is the swimming reflex, which can be elicited during the first 5 months if the child is held over water. When performing locomotor reflexes, an organized flexion and extension of both the upper and lower limbs can be seen (*Fig. 4*).

Having considered reflexes, we now need to consider the processes that underpin further motor development in infants. The role of reflexes in development is an area that has caused the greatest conflict between theories of motor development.

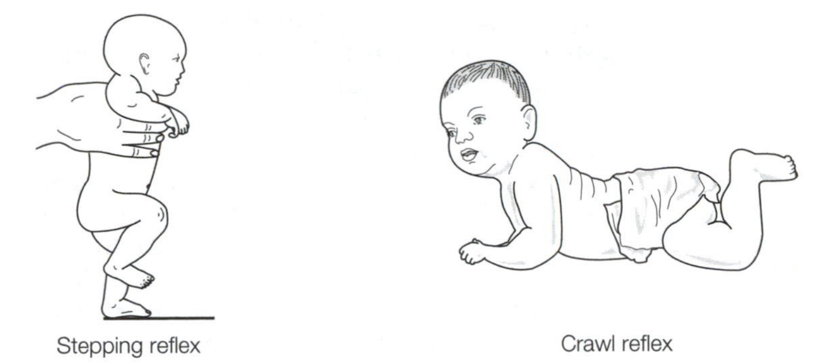

Fig. 4. Locomotive reflexes.

The process underlying development

Maturationalists stated that motor development was a linear stage-like progression, with the infant passing through a sequence in which there was an increase in more functional behavior. Motor development was considered to follow an orderly sequence allowing us to describe progressions of change (Keogh and Sugden, 1985). During these progressions a number of achievements are reached that represent the general pattern of change; these landmark achievements are conceived as 'milestones'. As a result progressions of change for locomotion, postural control and manual control were compiled from the recorded observations of Shirley (1931), Gesell and Ames, (1940) and McGraw (1946). These descriptions of motor milestones and stages were incorporated into textbooks and their age-norms became the bases of widely used assessment tools. It was assumed that the achievement of milestones reflected changes in the CNS. Thelen (1995) notes:

> . . . these early workers concluded that the regularities they saw as motor skills reflected regularities in brain maturation, a genetically driven process common to all infants. (p.79)

More specifically, McGraw noted that the achievement of milestones was due to maturation of the cortex. It was proposed that as the cortex matured it progressively inhibited the limited stereotyped subcortical output seen as reflexes, and began to take control over the neuromuscular system. McGraw consistently described a multidetermined, gradual process, with the structure, in this case the tissues of the brain, fundamentally involved in any developmental change. However, it soon became clear that motor developmentalists were unsatisfied with the maturational account of motor development and the next part of this section considers the weakness of this approach to development.

Deficiencies of maturationalist explanations

Thelen and Ulrich (1991) expressed their concerns about the deficiencies of the neuromaturational account of motor development, stating that the central problem of this perspective is that it ignores richness inherent in developmental behavior arising from many subsystems and processes. Indeed, if we accept the maturationalist perspective for the development of motor behavior, how do we explain the several dramatic transitions that are usually observed during development? For example, during the development of human upright locomotion at about 2 months of age there is an apparent 'disappearance' of the coordinated step-like movements performed by newborns when they are held erect. This behavior is then noted again during the second half of the first year of life. Thelen and Fisher (1982) noted that in the first two or three months, the stepping reflex disappears at a time when infants gain weight rapidly. This weight gain is mostly due to subcutaneous fat rather than muscle tissue, and as a result their limbs get heavier but not proportionately stronger. As a result, Thelen and Fisher (1982) speculated that the disappearance of the reflex could arise from the increasingly heavy legs and a biomechanically demanding posture, not through the voluntary cortical centers inhibiting the reflex. Further research by Thelen *et al.* (1984) substantiates this claim. Three-month-old infants who did not normally step performed stepping movements when their legs were submerged in water and consequently their body weight was supported. Conversely, the addition of small weights to the legs, thus mimicking the increase in weight observed during this time, suppressed the stepping movements. Therefore very small changes in the infants' weight or environmental context shifted the path of transition believed to be the inevitable consequence of brain maturation.

Further work by Thelen (1986) questions the ontogenetic origins of stepping behavior. While examining the second transition of upright locomotion, that of the 'reappearance' of stepping movements, it was found that infants would treadmill step at 7 months of age when supported on their feet. Normally this transition is late in the first year, yet simple mechanical manipulations reveal a different developmental course. Even more remarkable were infants' abilities to make functionally appropriate corrections to the bilateral coordination of their treadmill stepping (Thelen *et al.*, 1987a). Therefore it seems that motor behavior is composed of many subcomponents; not only are these components distinct, they also develop at different rates and all behaviors are entirely context dependent (Thelen and Smith, 1994). Thelen (1995) notes that under specific experimental conditions certain aspects of functional activity that are normally hidden can be elicited in advance of the fully formed action. This was clearly demonstrated by Thelen and Ulrich (1991) who conducted an experiment in which babies as young as one month old were held supported under the arms, so that their legs rested on a small, motorized treadmill. Subsequent coordinated alternating stepping movements were elicited in the infants. These stepping movements were seen to share many of the kinematic patterns observed in adult walking. Furthermore, infants were able to maintain excellent alternation of their legs on a split-belt treadmill, whereby one leg was driven on a belt at twice the speed of the other leg. Thus the authors also noted that treadmill stepping is sensitive to external conditions such as speed and direction of the belt. Thelen and Ulrich (1991) comment that treadmill stepping is truly a hidden skill, because without the facilitating effect of the treadmill, such patterns are not seen in babies until they begin to walk independently. Thus treadmill stepping is not just a simple reflex, but a complex, perceptual-motor pathway whereby the dynamic stretch of the legs provides both energetic and informational components that allow the complex pattern of stepping to emerge.

Research Highlight: Esther Thelen

Any section on motor development should take into account the work of Esther Thelen. Thelen, with a variety of colleagues, conducted ground-breaking research into motor development. Her work both challenged and extended out knowledge of how infants learn to reach, walk and accomplish tasks in a variety of contexts. Researchers used to believe that infants reached and walked when parts of the brain responsible for these activities 'matured'. The work of Thelen has shown that these accomplishments reflect a complex interplay of factors that includes infants' changing bodies, the external environment, and their developing brains and nervous systems. Her work has examined stepping in infants (Thelen and Fisher, 1982; Thelen *et al.*, 1984; Thelen 1986; Thelen and Ulrich 1991; Thelen and Smith 1994; Thelen, 1995) and she was also the first to identify a range of spontaneous movements produced by infants. Spontaneous movements will be considered in more detail later in this section, but the following are landmark studies.

- Thelen, E. (1979) Rhythmical stereotypies in normal human infants. *Anim Behav* **27**, 699–715.
- Thelen, E. (1985) Developmental origins of motor coordination: leg movements in human infants. *Dev Psychobiol* **18**, 1–22.
- Thelen, E. and Fisher, D.M. (1983a) From spontaneous to instrumental behavior: Kinematic analysis of movement changes during very early learning. *Child Dev* **54**, 129–140.
- Thelen, E. and Fisher, D.M. (1983b) The organization of spontaneous leg movements in newborn infants. *J Mot Behav* **15**, 353–377.

Cognitive approaches to development

Before discussing the development from a dynamical systems perspective we must consider cognitive approaches to development. The 1970s was a period of extensive research in the field of motor control. During this time cognitive approaches to control dominated and this had an influence on the field of motor development. The most influential work in the area of cognitive development was work by psychologist Jean Piaget (1896–1980). His theory provided many concepts in the field of developmental psychology that dominated for many years. Like other theories of motor control in the 1970s, his theory was concerned with the emergence and acquisition of schemata. Piaget described cognitive development in terms of schemes that children progressed through in four stages. These stages included the following:

- *sensorimotor stage*: years 0–2
- *preoperational stage*: years 2–7
- *concrete operational stage*: years 7–11
- *formal operational stage*: years 11 to adulthood.

In the sensorimotor phase, infants form and understand relationships between their bodies and the environment. Piaget claimed that children are born with little ability and without knowledge of reasoning or representation. During the first few months of life children develop schemas that enable them to organize and interpret information. He believed that primary reflexes formed early schemas with the infant having little else to draw on. In the preoperational stage the child begins to make relationships between schema and continues to explore the environment. The concrete operational stage is the third stage in Piaget's theory. Once this stage is reached the infant begins to make logical connections and is able to organize thoughts coherently. The final stage in Piaget's theory is the formal operational stage. The child now has the ability to formulate hypotheses and can systematically test them in a variety of contexts. Piaget stated that there are four interrelated factors that allow movement from one stage to another. These factors include maturation, experience, social interaction and equilibration. Equilibration occurs when the child is able to bring together maturation, experience and social interaction in order to build a mental schema. Piaget's work was very influential but it must be remembered that elements of his theory have now been disproved.

For example, it is clear that infants are capable of movements that are not reflex driven (Von Hofsten 1982; Thelen *et al.*, 1987b).

The contributions of Bernstein

We have established that the maturational perspective to motor development cannot fully explain motor development, as it provides no mechanism for understanding change or how new behaviors occur. Bernstein (1967) was the first to examine in detail the problems of the brain–behavior causal link and the problem of complexity. This work led directly to dynamic perspectives on motor coordination and developmental theory. Bernstein defines movement in terms of coordination and the cooperative interaction of many body parts and processes to produce a unified outcome (Turvey, 1990). Movements occur according to Bernstein because of an imbalance of forces caused by changes in muscle tension, thus rejecting the idea that a movement reflects a one-to-one relationship between the neural codes, precise firing of motor neurons and the actual movement pattern. In other words, it is impossible to map directly muscle activity patterns measured either in the CNS or at the periphery to the actual trajectory of the movement. Bernstein recognized that movements could come about from a variety of underlying muscle contractions, indicating that movements are not preprogramed in detail but at an abstract level are refined by the task. Indeed once a decision to move has been made, the subsystems and components that actually produce limb trajectory are softly assembled (Kugler and Turvey, 1987). Therefore during the execution of movement the device to be controlled presents the CNS with a continuously changing biodynamic challenge. As Thelen and Ulrich (1991) note:

> This type of organisation allows the system more flexibility to meet the demands of a task within a continually changing environment, while maintaining a movement category suited to the goal in time. (p.81)

In light of the work of Bernstein, the concept of waiting for the brain to mature and then execute the brain's demands is clearly untenable. These observations cast doubt on any explanation that relies heavily on simple nervous system maturation as the process by which infants develop motor coordination. Unlike the maturational and information-processing perspectives, the dynamical systems approach suggests coordinated behavior is 'softly assembled' rather than 'hard wired', in that there is no preset plan that exists in order to produce functional skills. In Bernstein's (1967) terms, these skills 'develop and involute' (p.180). The product of this soft assembly is that we have greater flexibility when performing functional skills. For infants as well as adults, movements are always the product not only of the CNS, but also of biomechanical and energetic properties of the body in conjunction with environmental support and the specific demands of the task. Thelen (1995) notes that the relationship between these components is not hierarchical, in that the brain commands and the body responds. Rather, it is profoundly distributed 'heterarchical, self-organizing and non-linear. Every movement is unique: every solution is fluid and flexible' (p.81).

The traditional explanation of development was single-causal: maturation of the voluntary cortical centers first inhibits subcortical or reflexive movements and then facilitates them under a different and higher level of control (McGraw, 1945). This long-accepted explanation came into question due to the work of Bernstein

(1967) and the work of Thelen (1979) and Thelen and Fisher (1983a). At this point it is appropriate to examine spontaneous movements before considering in depth the contribution of dynamical systems approaches to development.

Spontaneous movements

Thelen (1979), Thelen and Fisher (1983a) and Piek and Carmen (1994) have argued that a self-organizing group of infant behaviors may be the fundamental structure underlying the construction of voluntary movement. These movements are referred to as spontaneous movements and include kicking, rocking, scratching, waving, bouncing, banging and swaying and are performed by babies in the first year of life. Thelen (1979) was the first to produce a comprehensive classification of early spontaneous movements from infants aged 4 weeks to 12 months. Twenty infants were observed every 2 weeks, and a total of 47 different types of rhythmical stereotypies (spontaneous movements) were identified, including body rolls, arm waves, rocking and kicking. Bouts of movement were recorded, and defined as:

> a movement of parts of the body or the whole body that was repeated in the same form at least three times at regular short intervals of about a second or less. (Thelen, 1979, p.700)

Thelen (1979) noted the mean number of bouts increased between the ages of 4 and 54 weeks. Piek and Carmen (1994) carried out a similar observational analysis to Thelen (1979), but used a less-stringent definition where bouts could be single movements (rather than three or more). Video analysis was used in this cross-sectional study of 50 infants ranging from 1–50 weeks of age. As a result, 53 different patterns of spontaneous activity were identified. In contrast to Thelen (1979), the frequency of spontaneous movements increased between 1 and 20 weeks of age and then reduced in frequency as the infant became older (*Fig. 5*). Similar results were again recorded by Piek *et al.* (2001). The findings by Piek and Carmen (1994) and Piek *et al.* (2001) demonstrate that a measure such as frequency is dependent upon the criteria set for movement to be measured.

Spontaneous movements are not reflexes or involuntary movements, and it appeared that they were not used in a goal-directed sense. Therefore, it was thought that they serve little or no purpose. However, the work of Thelen and Fisher (1983a, b) has shown that the spontaneous movements of newborns are actually organized, rhythmic and stereotyped at the kinematic level. Initially these spontaneous movements are subject to temporal organization with the limbs extending and flexing as a single unit (Thelen and Fisher, 1983a). By the middle of the first year this tight coordination diminishes and the infants move their joints more individually, which is necessary to establish a more adaptive pattern of coordination and control. Thelen and Fisher (1983a) concluded that the development of self-organizing systems occurs as a result of assembling and disassembling system linkages by spontaneous movements. Such reorganization of the movement results in an initial decline in the motor control but this is followed by a period of reforming which enhances development (Mounod and Vinter, 1981). It appears that spontaneous movements are precursors to later development in the same way that reflexes provide the basis for later voluntary control of movement. Spontaneous movements are seen as transition behaviors between early coordinated activity and the development of full voluntary control (Thelen and Fisher, 1983a).

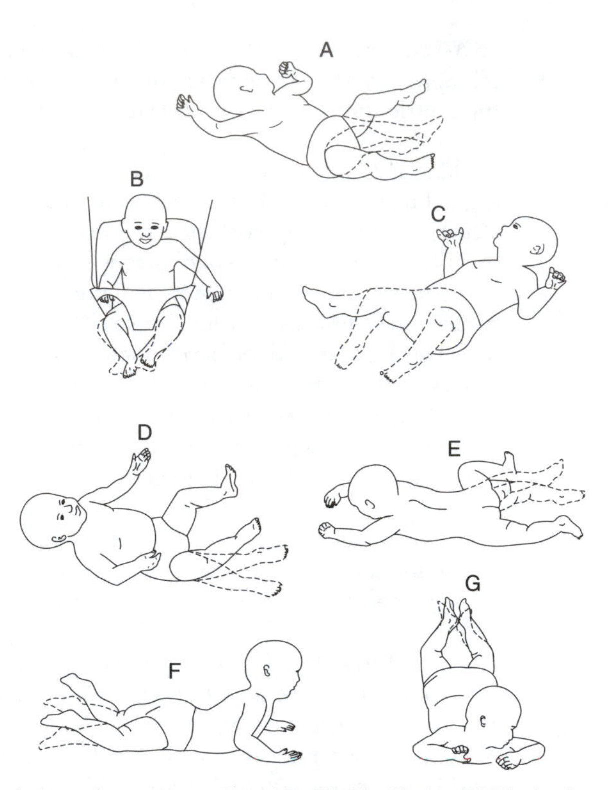

Fig. 5. Spontaneous leg movements as identified by Thelen (1979). A, alternate leg kicking, supine position; B, foot rubbing; C, single leg kicking, supine position; D, single leg kicking with leg rotated outward; E, single leg kicking prone position; F, both legs simultaneous kicking; G, both legs simultaneous kicking back arched. Reproduced with permission from Thelen (1979), Figure 2.

Thelen (1985) also investigated the relationship between the two limbs when kicking spontaneously. This was characterized by an alternating leg kicking in the newborn infant up to 1 month of age, followed by a period of primarily single-leg kicking until ~4 months of age. Following this there appears to be a coupling of the limbs resulting in simultaneous bilateral kicking. This latter finding was supported by Piek and Carmen (1994). The work of Thelen (1985) and Piek and Carmen (1994) suggest that the coupling of the limbs 'tightens' with maturation.

The work of Thelen and Fisher (1983a, b) also showed that by the middle of the first year the tight coordination between and within joints of the limbs diminishes and the infants move their joints more individually, which is necessary to establish a more-adaptive pattern of coordination and control. Thelen and Fisher (1983a, b) concluded that the development of self-organizing occurs as a result of assembling and disassembling system linkages by spontaneous movements. This allows the infant to explore the intrinsic dynamics of the limbs, and from this more complex patterns of motor patterns emerge (Turvey and Fitzpatrick, 1993).

Thelen and Fisher (1983a) measured the spontaneous movements seen in young babies when supine, and noted that the kinematics of these movements resembled spatial and temporal components of mature walking patterns. This work suggests that new movements seen in infants are not simply the result of autonomous brain function as the neuromaturational perspective assumes. Thus McGraw's single-causal explanation of development does not seem possible. Thelen and Fisher (1983a) indicated that movements made when supine and when walking were spatially and temporally the same, but made in different positions. It thus seems unlikely that the cortex would inhibit movements in one posture but not in another. It appears as though spontaneous movements are precursors to later development and are seen as transition behaviors between early coordinated activity and the development of full voluntary control (Thelen and Fisher, 1983a).

A dynamical systems approach to motor development

More recent theories of motor development utilize principles of dynamical systems, which attempt to explain how complex systems, for example developing humans, develop functionally appropriate behavior. Bernstein (1967) asked how the nervous system coordinates all the body parts to produce functional movement given the large number of degrees of freedom in the neuromuscular system. The problem Bernstein examined is the same problem that theorists must investigate to understand the generation of new behavior.

The key difference between early neuromaturational theories and the dynamical systems approach is that the former suggests that change comes from a pre-existing genetic code, while the latter views change as a self-organizing process. It is the task that motivates the individual, not simply just a case of maturation. The task interacts with the child's resources thus producing a driving force needed to elicit change (Sugden and Wright, 1998). In self-organization, the system selects or is attracted to one preferred configuration out of many possible states, but behavioral variability is an essential precursor.

Dynamical stability and attractors

When systems self-organize under the influence of an order parameter, they 'settle onto' one or a few modes of behavior that the system prefers over all possible modes (Thelen and Smith, 1994). From a dynamical systems perspective this behavior mode is an attractor state, in that the system prefers a certain topology in its state space. The state space of a dynamical system is an abstract construct of a space whose coordinates define the components of the system, the degrees of freedom of the system's behavior and the coordinates that vary with changes in context (Thelen and Smith, 1994).

Several characteristics of attractors are particularly important as a construct in development. Complex systems seek preferred behavioral modes as a function of the interactions of their internal components and their sensitivity to external conditions. The attractor regime is only determined as the system is assembled through the slaving of its order parameter, the order parameter acting to constrain or compress the degrees of freedom available. The result is that despite the potential for chaos, complex systems become highly ordered and coordinated when in an attractor state, and attractor states possess varying degrees of stability and instability.

Thus dynamic change can be seen in dynamic terms as a series of stability, instability and phase shifts in the attractor landscape, reflecting the probability that a pattern will emerge under specific constraints. When the system is stable it is performing the preferred patterns, and change is brought about by a disruption of this stable position. Once new configurations are possible and discovered, they must be progressively tuned to become efficient, accurate and smooth. Thus, Thelen (1995) believes that for any particular task, a dynamic view predicts an initial high variability in configurations representing an exploration stage, a narrowing of possible states to a few patterns, and a progressive stability as the patterns become more practiced and reliable. Clearly, theory predicts that times of instability are essential to the system to allow flexibility and to select adaptive activities.

The work of Clark and Phillips (1993) examined stability and instability using a dynamical systems strategy by analyzing collective variables for the leg's segmental motion over a one-year period in infants. Clark and Phillips noted that from the very onset of independent walking, the infant is attracted by the same dynamic solution as the adult, but, as predicted by the dynamics of the system in transition, the solution is unstable. In addition, the infants refused to walk when a weight was added to their ankle. However, within 3 months the system had stabilized. Similarly this occurred with the thigh and shank, with an adult-like pattern appearing after 3 months of walking. Their results indicated that certain aspects of intralimb coordination became increasingly stable with development, and that overall, the development of upright locomotion can be modeled in terms of the behavior of a dynamic system.

A number of authors have used Waddington's (1954) epigenetic landscape to reflect how development takes place (Thelen, 1995). Thelen (1995) notes that as a ball runs through the landscape, a deep narrow fissure is characterized by stability while at the same time having a limited number of options, whereas the landscape with no fissures is essentially unstable and has many options. A key point is that the landscape for a dynamic system only ever offers temporary stability at most (Kauffman, 1993, 1995). Dynamic systems are more or less likely to be stable depending on the nature of internal and external constraints pressurizing system stability at any one time. The type of stability that characterizes dynamical movement systems therefore is 'dynamic stability' (Kelso, 1981). Therefore at any instant a dynamical movement system is either temporarily assembled into an attractor or is flowing to an attractor. Hilly or broadly flat regions of the attractor landscape are unstable and represent more complex skills, needing a small external perturbation to change it (*Fig. 6A*). The parameter or external shift that the system is sensitive to, causing the system to move onto another attractor, is a control parameter. It assembles the system onto one or another attractor regime given the states of the system subcomponents. The less-hilly regions of the landscape offer more stability to a dynamical movement than others. If the hollows are small and deep with relatively steep sides this offers a stable region for the system to settle in (*Fig. 6B*). So when looking at *Fig. 6*, the ball in A is more unstable as it could move in any direction. However, the ball in B has fewer options and is therefore more stable.

The most successful application of the above dynamical systems principles to human motor development concerns the development of cyclic behavior, in particular, the development of walking and running. The data of Thelen and

Fig. 6. Stable and unstable attractors. A, a very unstable attractor; B, a very stable attractor. Adapted from Thelen and Smith (1994), p.60.

Fisher (1983a) make more sense from a dynamic view. Thelen *et al.* (1987a) suggest that the early kicking movement noted by Thelen and Fisher (1983a) were examples of oscillatory movement in human limbs; thus patterns of interlimb coordination reflect various coupling mechanisms between single-leg oscillators. However, the infant must be able to dissolve such synergies for more specific and appropriate combinations of limb control, in order to assemble new task-specific patterns within and between joints. Jensen *et al.* (1995) found that infants show the progressive ability to move their leg joints independently rather than as a whole unit, so allowing finer control of forces through the knee and hip. This decoupling of the two legs occurs by freeing degrees of freedom, providing the child with greater flexibility to explore and discover new patterns.

Exploration and selection

The roles of exploration and selection in finding solutions to new task demands are strongly emphasized using a dynamical systems explanation of motor development. An infant modifies current movement dynamics to assemble adaptive patterns depending upon the task. This change is best seen when infants are given a novel task, task novelty is extremely important when the aim is to demonstrate a process where the outcome could not be anticipated by phylogeny or neural codes (*Fig. 7*). Goldfield *et al.* (1993) monitored the development of an infant learning to use a 'baby bouncer' and found that they adjusted their kicking to gain optimal bounce. The researchers predicted that two processes would occur when the infants, 6 months of age, were placed in a baby bouncer. Firstly, there would be an 'assembly' of the action system with

Fig. 7. Young infants are keen to explore their environment.

low-dimensional dynamics as the child explored the situation. This is followed by a tuning of the system to refine and adapt movements to the particular condition. The early assembly, as was predicted, was characterized by sporadic irregular kicking followed by more periodic kicking as the tuning took place. The third prediction suggested that once the bouncing was optimized (as a stable attractor) there would be a decrease in variability and an increase in amplitude, phase locking of kicking, bouncing and phase resetting as a result of the perturbation. Lastly, it was predicted that the child would be able to adapt to changing conditions, such as a new bouncer with different dynamics.

Infants in this study moved to an optimized attractor state where they got maximal bounce for minimal energy input. Similar results have also been documented more recently by Goldfield (2001) with respect to oral-respiratory coordination and the production of speech. Goldfield (2000) states that the results yielded by the production of speech complement those found in 1993 on locomotion in infants. It was suggested that just as infants of 6–9 months of age seem to explore and subsequently discover when to kick and take advantage of the potential energy stored in the legs, the infant discovers when to produce a speech segment and to take advantage of the potential energy stored in the lungs produced by canonical babbling or replicated babbling (Oller, 1980). The data on locomotion and speech suggest the possibility that a general competence that emerges in the first year is the ability to tap into sources of potential energy stored in the body and sustain long sequences of action (Goldfield, 2000).

More research that complements the work of Goldfield *et al.* (1993) is that of Thelen (1994). Thelen (1994) attached one of a baby's legs to an overhead mobile with a ribbon tied around the ankle, such that the infant's spontaneous kicking made the mobile move. Interestingly, the babies learnt that the mobile moved and created noise in direct relation to how frequently and vigorously they kicked. In addition, Thelen also investigated whether the infants could learn a new pattern of interlimb coordination. Thelen (1994) tethered the baby's legs together to inhibit the use of independent kicking and to see if she could experimentally induce the infants to use a simultaneous in-phase pattern, therefore investigating whether the tethering could disrupt the stability of old attractors, allowing the infant to explore and discover more efficient leg-coupling attractors. Indeed Thelen (1994) found just what she had hypothesized. These results also indicate the effect of intentionality as a control parameter, as a new pattern of coordination arises from task demands placed upon existing coordinative tendencies (Corbetta and Thelen, 1996).

The neuronal group selection theory

Sporns and Edelman (1993) propose the neuronal group selection theory (NGST) as providing the basis for most development. Sporns and Edelman refute the idea that development evolves from genetic instructions, and instead propose that through periods of instability (where many options to solve the motor problem are experienced) and stability (where far fewer options are used to solve a motor problem), humans select solutions that strengthen certain connections or groups of neurons through use. Development must initially involve a basic movement repertoire and the ability to sense the effect of the movement on the environment, and it must include a mechanism by which those movements are selected, satisfying both environmental demands and internal constraints. Selection is therefore a key aspect of change, and successive selection will involve progressive modifications of a given movement repertoire. As noted by Sugden and Wright (1998), this alleviates the need to solve a problem by computations or processing means, and replaces it with an organism selecting purposeful movement synergies from a wide range, that have adaptively developed to solve environmental and biomechanical problems.

Translating concepts of the NGST provided by Sporns and Edelman (1993) to the domain of human motor development results in developmental progress with two specific phases of variability (Hadders-Algra, 2000). Hadders-Algra (2000) identifies two types of movement variability as movements develop, namely 'primary' and 'secondary' variability. Primary variability occurs early in fetal life and continues into infancy. Properties of primary variability are very well illustrated by spontaneous movements, the rich variation and complexity of these reflecting the explorative activity of a widely distributed (sub)cortical network, the primary neuronal group. These spontaneous movements gradually disappear giving way to more goal-directed movements. The neural systems dedicated to these more specific behaviors explore (during the phase of primary variability) all the motor processes available for that specific behavior, which gradually results in the selection of the most efficient movement pattern.

Through the process of 'selection', the infant moves into the next stage of variability called 'secondary' or 'adaptive' variability (Hadders-Algra, 2000). This selection process is based upon Edelman's adaptive value, where the movements that are more successful in accomplishing the task result in the strengthening of the appropriate synaptic connections (Sporns and Edelman, 1993). According to Sporns and Edelman (1993), 'efficiency' is determined through an adaptive value system that identifies the value of particular movements. They suggest that different value systems are available for functionally interrelated tasks, and that these systems interact. This selection stage results in a temporary reduction in variability as the infant selects the movement patterns that appear most efficient. Hadders-Algra (2000) argues that the selection stage is functionally specific, occurring at different developmental ages for milestone such as reaching, crawling and walking. Once the selection stage is complete, there is again increased variability, termed 'secondary variability.' Hadders-Algra (2000) argues that when a child is given a novel task to perform, considerable practice is needed to reduce secondary variability in order to produce the most efficient motor solution. The ability to adapt each movement exactly and efficiently to task-specific conditions or in the absence of tight constraints, by generating multiple solutions or strategies for a single motor task, characterizes the mature situation of a healthy adult, culminating from variability, exploration and selection as proposed by the NGST.

Hence, practice and experience are considered essential aspects of development of motor skills. The NGST highlights the importance of variability in the development of movement, and emphasizes the need for a variable, diverse repertoire of movement patterns to be available early in development for the selection of an appropriate solution to a given movement problem (Thelen and Smith, 1994). Sporns and Edelman (1993) argue that movements emerge as a result of the interplay between the neuromuscular system and the environment, emphasizing the interrelationship between perception and action. They suggest that what is missing from a dynamical systems account of motor development is the identification of the specific neural mechanisms that account for development from an emergent rather than a prescribed perspective.

Perception and action

A central theme in the dynamic systems approach to motor development is the inseparable coupling of perception and action in the generation and improvement of new skills. The action–perception coupling research has been directly inspired by theories of E.J. Gibson (1988) and J.J. Gibson (1979) on perception and perceptual development. These studies demonstrated that, from the beginning, infants are continually coordinating their movements with concurrent perceptual information to learn how to maintain balance, reach for objects in space and locomote across various surfaces.

The ecological view of perception and action is based on direct perception of the environment. Developmentalists from this perspective believe that we also directly perceive what objects and surfaces in the environment permit us, given our own capabilities; that is, we perceive affordances. Gibson (1988) propounds that an affordance is the reciprocal relation between the actor and the environment that is necessary to perform functional activities. However, the developmental question is how these relationships are first acquired and subsequently how do infants learn that they can perform certain actions within their environment?

Adolph *et al.* (1993) present evidence of infants' growing ability to detect affordances. Crawlers and toddlers were presented with a sloping surface of various degrees of steepness. Adolph and colleagues observed patterns of locomotion over such surfaces in order to answer two research questions: would the children know when they could successfully go up or down the slope without falling, and would they be able to adjust their patterns of locomotion to the steepness of the slope? Ascending the slope posed no problems for either the toddlers or crawlers, and both groups even tried the steepest of slopes without hesitation. However, when descending, as the steepness increased, the more the toddlers became wary and in some cases often refused to go. This suggested that the toddlers appeared to understand something about their locomotor abilities and whether they fitted the task. In contrast, the crawlers plunged downhill suggesting that they did not appear to perceive such a fit between ability and task (*Fig. 8D*).

Although the specific factors contributing to this difference are unclear, the authors specify that it is likely that the process involved the infant's own continuing exploration of their action capabilities in relation to the slopes, and learning and remembering about the consequences of their activities. Adolph *et al.* (1998) also point to the fact that the infant crawlers are less attentive to their own postural stability than toddlers who are possibly more focused due to the demands of balancing on two legs. Stoffregen *et al.* (1997) undertook a study to determine whether young children with only a few weeks' standing experience were also able to maintain and adapt their standing posture according to the constraints of different support surfaces. The study assessed if children would use manual control (holding a pole) as a strategy for maintaining postural control and looked at these adaptations in the absence of imposed perturbations. The children were seen to exhibit remarkable adaptive postural control in relation to the varying surfaces, these being a high-friction surface, a soft mattress, a rigid low-friction surface and a narrow wooden beam. The children used complex movements to maintain postural control including movements of the hips (not previously thought possible until 3 or 4 years of age) and ankles, and use of arms and legs. The children used a wide variety of coordination modes in the maintenance of their stance. They utilized the poles in a surface-specific manner, indicating, the authors believe, perception of and adaptive response to at least some of the constraints of the surfaces to their postural control.

The primary thrust of research from a Gibsonian approach has been to under-

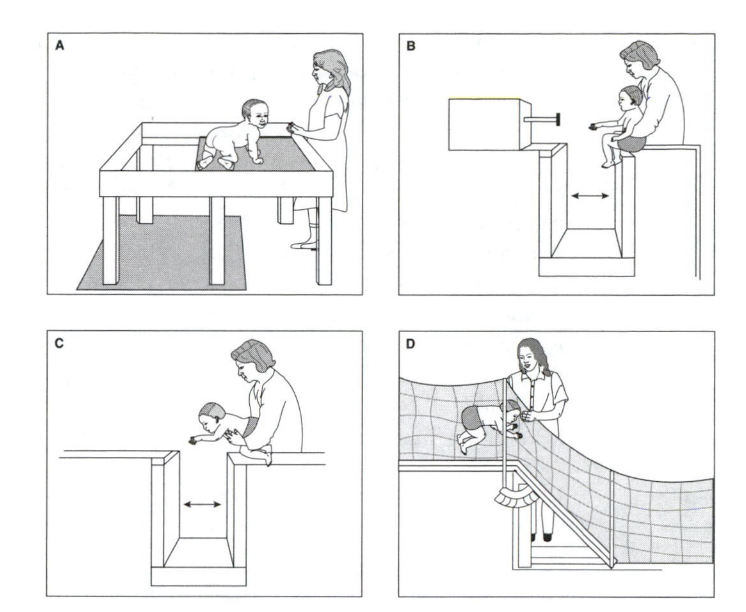

Fig. 8. In a series of studies Adolph (2000) and Adolph et al. *(1993) have examined infants' responses to differing locomotor contexts. Experience appears to influence performance, with infants with 6 weeks' experience of crawling avoiding the gap, but those with only 2 weeks' experience attempting to cross. In C and D the infant was tested in either a sitting or crawling posture. In both conditions the infant had to reach across a gap (of varying width) and get a toy, the infants had 15 weeks' sitting experience and 6 weeks' crawling experience. In the sitting posture the infants matched their movements to the probability of falling but when in the crawling position they made the wrong decision. Taken from Adolph (2000) and Adolph* et al. *(1993).*

stand how perception guides action. Thelen (1995) points out that motor activities are particularly useful because they provide the means for exploring the world and for learning about its properties. During development each motor milestone opens new opportunities for perceptual discovery. Thelen (1995) believes that it may be continual bombardment of sensory information that creates developmental change as infants learn to act in social and physical worlds. The key elements are the dynamic processes of exploration and selection: the ability to generate behavior that provides a variety of perceptual-motor experiences and then the differential retention of those correlated actions that enable the infant to function in the world.

Conclusion

Explaining motor development goes beyond mere descriptions and involves an examination of the processes that drive changes seen in children. The maturation and information-processing approaches to motor development have dominated explanations of change for many years, but did not provide satisfactory explanations for performance or development of motor behavior. Information-processing theories have provided insightful ways with which to examine and remediate motor difficulties. Descriptions of what children need to know as well as do when performing a motor skill aid the development of intervention programs (Sugden and Wright, 1998). However, more recently the dynamical systems approach has been the driving force behind renewed interest of motor development in several ways. Sugden and Wright (1998) comment on the attractiveness of the dynamical systems approach for analysis of the nature of coordination disorders, assessment in context and principles for intervention. The more recent views of motor development in the 1990s stress the role of exploration and selection of solutions to demands of novel tasks. These new views contrast with the traditional maturational accounts of motor development by proposing that the so-called 'phylogenetic skills' such as reaching and grasping and crawling and walking are in fact learnt by adapting current dynamics through exploration and selection of possible movements to fit the new task (Thelen, 1995). Consequently there are a number of ways in which motor development can be illustrated using maturational theory and/or dynamical systems principles, enabling both a behavioral description and a full description of the process to be obtained. In the next section the development of fundamental motor skills such as posture, reaching/grasping, catching and locomotion are described from birth through to adolescence.

Further reading

Jackson, S.L. and Healey, J. (2000) Interfacing computerized biomechanical analysis in monitoring the motor development of children. *Percept Mot Skills* **91**, 999–1008.

Okely, A.D. and Booth, M.L. (2004) Mastery of fundamental movement skills among children in New South Wales: prevalence and sociodemographic distribution. *J Sci Med Sport* **7**, 358–372.

Piek, J.P. (2006) *Infant Motor Development*. Human Kinetics, Champaign, IL.

Smith, L.B. and Thelen, E. (2003) Development as a dynamic system. *Trends Cogn Sci* **8**, 343–348.

Spencer, J.P. and Thelen, E. (1999) A multimuscle state analysis of adult motor learning. *Brain Res Exp Brain Res* **128**, 505–516.

References

Adolph, K.E. (2000) Specificity of learning: why infants fall over a veritable cliff. *Psychol Sci* **11**, 290–295.

Adolph, K.E., Eppler, M.A. and Gibson, E.J. (1993) Crawling versus walking infants' perception of affordances for locomotion over sloping surfaces. *Child Dev* **64**, 1158–1174.

Adolph, K.E., Vereijken, B. and Denny, M.A. (1998) Learning to crawl. *Child Dev* **69**, 1299–1330.

Bernstein, N. (1967) *Coordination and Regulation of Movements*. Pergamon Press, New York.

Clark, J.E. and Phillips, S.J. (1993) A longitudinal study of the intralimb coordination in the first year of independent walking. *Child Dev* **64**, 1143–1157.

Corbetta, D. and Thelen, E. (1996) The developmental origins of bimanual coordination: a dynamic perspective. *J Exp Psychol Hum Percept Perform* **22**, 502–522.

Gallahue, D.L. (1982) *Understanding Motor Development in Children*. John Wiley and Sons, New York.

Gessell, A. (1939) Reciprocal interweaving in neuromotor development. *J Comp Psychol* **70**, 161–180.

Gesell, A. and Ames, L.B. (1940) The ontogenetic organisation of prone behaviour in human infancy. *J Genet Psychol* **56**, 247–263.

Gibson, E.J. (1988) Exploratory behavior in the development of perceiving, acting and acquiring knowledge. *Annu Rev Psychol* **39**, 1–41.

Gibson, J.J. (1979) *The Ecological Approach to Visual Perception*. Houghton-Mifflin, Boston.

Goldfield, E.C. (2000) Development of infant action systems and exploratory activity: a tribute to Edward S. Reed. *Ecol Psychol* **12**, 303–318.

Goldfield, E.C. (2001) Production of vowel sounds by term and pre-term infants: exploration of vocal tract resonances. *Infant Behav Dev* **23**, 421–439.

Goldfield, E.C., Kay, B.A. and Warren, W.H. Jr. (1993) Infant bouncing: the assembly and tuning of action systems. *Child Dev* **64**, 1128–1142.

Hadders-Algra, M. (2000) The neuronal group selection theory: a framework to explain variation in normal motor development. *Dev Med Child Neurol* **42**, 566–572.

Haywood, K.M. and Getchell, N. (2001) *Life Span Motor Development*, 3rd Edn. Human Kinetics, Champaign, IL.

Hellebrandt, F.A., Schade, M. and Carns, M.L. (1962) Methods of evoking the tonic neck reflexes in normal human subjects. *Am J Phys Med* **41**, 90–139.

Jensen, J.L., Thelen, E., Ulrich, B.D., Schneider, K. and Zernicke, R.F. (1995) Adaptive dynamics of the leg movement patterns of human infants: III. Age-related differences in limb control. *J Mot Behav* **27**, 366–374.

Kauffman, S.A. (1993) *The Origins of Order: self-organisation and selection in evolution.* Prentice Hall, New York.

Kauffman, S.A. (1995) *At Home in the Universe: the search for laws of complexity.* Viking, London.

Kelso, J.A.S. (1981) Contrasting perspectives on order and regulation in movement. In: Long, J. and Baddeley, A. (eds) *Attention and Performance* IX. MIT Press, Cambridge, MA, pp.437–457.

Kelso, J.A.S. (1982). *Human Motor Behavior: an introduction.* Erlbaum, Hillsdale, NJ.

Keogh, J.F. and Sugden, D.A. (1985) *Movement Skill Development.* Macmillan, New York.

Kugler, P.N and Turvey, M.T. (1987) *Information, Natural Law and the Self-assembly of Rhythmic Movement.* Erlbaum, Hillsdale, NJ.

McDonnel, P.M., Corkum, V.L. and Wilson, D.L. (1989) Patterns of movement in the first 6 months of life: new directions. *Can J Psychol* **43**, 320–339.

McGraw, M.B. (1932) From reflex to muscular control in the assumption of an erect posture and ambulation in the human infant. *Child Dev* **3**, 291–297.

McGraw, M.B. (1940) Neuromuscular development of the human infant as exemplified in the achievement of erect locomotion. *J Pediatr* **17**, 747–771.

McGraw, M.B. (1945) *The Neuromuscular Maturation of the Human Infant.* Columbia University Press, New York.

McGraw, M.B. (1946) Maturation of behavior. In: Carmichael, L. (ed) *Manual of Child Psychology.* Wiley, New York, pp.332–369.

Mounod, P. and Vinter, A. (1981) Representation and sensori-motor development. In: Butterworth, G. (ed) *Infancy and Epistemology.* Hassocks, London, pp. 96–156.

Oller, D.K. (1980) The emergence of speech sounds in infancy. In: Yeni-Komishian, G., Kavanagh, J.F. and Fergurson, F.C. (eds) *Child Phonology: 1. Production.* Academic Press, New York, pp.93–112.

Peiper, A. (1963) *Cerebral Function in Infancy and Childhood.* Consultant's Brown, New York.

Piek, J.P. and Carmen, R. (1994) Developmental profiles of spontaneous movements in infants. *Early Hum Dev* **39**, 109–216.

Piek, J.P., Rolston, C. and Gasson, N. (2001) The relationship between infant spontaneous leg movements and the asymmetrical tonic neck reflex in full term and low-risk preterm infants. In: van der Kamp, J., Ledebt, A., Savelsbergh, G. and Thelen, E. (eds) *Advances in Motor Development and Learning in Infancy*. Printpartners Ipskamp, Amsterdam, pp.69–72.

Prechtl, H.F.R. (1977) *The Neurological Examination of the Fullterm Newborn Infant. Clinics in developmental medicine*, No. 63. Mac Keith Press, London.

Schmidt, R.A. (1975) A schema theory of discrete motor skills learning. *Psychol Rev* **82**, 255–260.

Shirley, M.M. (1931) *The First Two Years: a study of twenty-five babies. Vol 1: Postural and locomotor development.* University of Minnesota Press, Minneapolis.

Sporns, O. and Edelman, G.M. (1993) Solving Bernstein's problem: a proposal for the development of coordinated movement by selection. *Child Dev* **64**, 960–981.

Stoffregen, T.A., Adolph, K., Thelen, E., Gorday, K.M. and Sheng, Y.Y. (1997) Toddlers' postural adaptations to different support surfaces. *J Mot Control* **1**, 119–137.

Sugden, D.A. and Wright, H.C. (1998) *Motor Coordination Disorders in Children. Developmental clinical psychology and psychiatry*, No. 39. Sage Publications, Thousand Oaks, CA.

Thelen, E. (1979) Rhythmical stereotypies in normal human infants. *Anim Behav* **27**, 699–715.

Thelen, E. (1985) Developmental origins of motor coordination: leg movements in human infants. *Dev Psychobiol* **18**, 1–22.

Thelen, E. (1986) Treadmill-elicited stepping in seven-month-old infants. *Child Dev* **57**, 1498–1506.

Thelen, E. (1994) Three-month-old infants can learn task-specific patterns of interlimb coordination. *Psychol Sci* **5**, 280–285.

Thelen, E. (1995) Motor development: a new synthesis. *Am Psychol* **50**, 79–95.

Thelen, E. and Fisher, D.M. (1982). Newborn stepping: an explanation for a disappearing reflex. *Dev Psychol* **18**, 760–775.

Thelen, E. and Fisher, D.M. (1983a) From spontaneous to instrumental behavior: Kinematic analysis of movement changes during very early learning. *Child Dev* **54**, 129–140.

Thelen, E. and Fisher, D.M. (1983b) The organization of spontaneous leg movements in newborn infants. *J Mot Behav* **15**, 353–377.

Thelen, E., Fisher, D.M. and Ridley-Johnson, R. (1984) The relationship between physical growth and a newborn reflex. *Infant Behav Dev* **7**, 479–493.

Thelen, E., Kelso, J.A.S. and Fogel, A.L. (1987b) Self-organizing systems and infant motor development. *Dev Rev* **7**, 39–65.

Thelen, E. and Smith, L.B. (1994) *A Dynamic Systems Approach to the Development of Cognition and Action*. MIT Press, Cambridge, MA.

Thelen, E. and Ulrich, B.D. (1991) Hidden skills: a dynamical systems analysis of treadmill stepping during the first year. *Monogr Soc Res Child Dev* **223**, 56.

Thelen, E., Ulrich, B.D. and Niles, D. (1987a) Bilateral coordination in human infants: stepping on a split-belt treadmill. *J Exp Psychol Hum Percept Perform* **13**, 405–410.

Turvey, M.T. (1990) Coordination. *Am Psychol* **45**, 938–953.

Turvey, M.T. and Fitzpatrick, P. (1993) Commentary: Development of perception-systems and general principles of pattern formation. *Child Dev* **64**, 1175–1190.

Twitchell, T.E. (1970) Reflex mechanisms and the development of prehension.

In: Connolly, K.J. (ed) *Mechanisms of Motor Skill Development*. Academic Press, London, pp. 25–45.

von Hofsten, C. (1982) Eye-hand coordination in the newborn. *Dev Psychol* **18**, 450–461.

von Hofsten, C. (1993) Studying the development of goal-directed behaviour. In: Kalverboer, A.F., Hopkins, B. and Geuze, R. (eds) *Motor Development in Early and Later Childhood: longitudinal approaches*. Cambridge University Press, Cambridge, pp.109–125.

Waddington, C.H. (1954) The integration of gene-controlled processes and its bearing on evolution. *Proceedings of the 9th International Congress of Genetics (Caryologia Supplement)*, **9**, 232–245.

Development of fundamental movement skills

Key Notes

Locotive skills —

Locomotive skills: Movement skills such as walking and running that enable us to move from one place to another.

Manual skills: Movement skills made with the hands such as grasping and catching.

Perturbation: A modification or change that is made to a task or the context in which the task is performed; often part of an experimental design.

Related topics: Theories of motor learning (H); Motor development (M); Implications for practice (K)

Fundamental movement skills begin to emerge before we are born, and rapid development takes place during the first nine years of life. Throughout life, movement skills continue to be acquired, developed and refined. The stages of development can best be considered as phases that the child passes through. It must be remembered that development is a continuum and that individuals develop at different rates. Therefore, when looking at developmental milestones it must be stressed that all ages are approximations and that different children will develop at different rates. This section will consider the development of the fundamental motor skills of posture, locomotion and manual skills. The starting point, however, will be to provide an overview of change from birth to puberty.

The first two years

During the first two years of life most children develop rapidly and make great progress in acquiring the basic skills of posture, locomotion and manipulation. Postural development is the key developmental change that takes place during these years. Postural development as we shall see later is a vital skill, as it is the precursor to locomotion and involves controlling the head and trunk while lying and sitting, eventually progressing to an upright stance. The development of upright locomotion is a key milestone, usually occurring around one year of age (Keogh and Sugden, 1985). Once the child can move independently around the environment, other skills can then be developed. Manipulation progresses steadily during the first two years of life. During this time reaching becomes more accurate and grasping, changing from a crude palmar grasp to the delicate pincer grip of the thumb and first finger. The child is also able to release objects with a high degree of control and is able to manipulate objects.

Two to seven years of age

During this period the child develops all of the basic locomotor skills and manual skills together with self-help skills. At the end of this period most children will be able to run, jump, hop, possibly skip, climb, throw, catch, kick, strike, dress themselves, wash themselves and perform everyday manipulation skills. These skills may be performed with some difficulty and some errors may well be made, but the child will be able to produce the basic movements of these skills. It is vital for children at this stage to develop a vocabulary of motor skills that will allow them to flexibly employ these skills in a variety of situations. All children should therefore be exposed to a wide range of movement skills in a variety of contexts.

Seven to puberty

At this period of motor development, the child rarely acquires new fundamental skills. During this phase the child is refining, honing and elaborating upon the skills they already possess. This is an important phase of developing as the child continues to explore the environment and learns to adapt and transfer skills into a variety of contexts. Skills such as interception and anticipation are developed during this period. Children are therefore able to learn to cross the road, intercept a moving object, and time their movements to coincide with those of others. Their perceptual and cognitive systems will not be as fully developed and children will need help with prediction and anticipation skills. The development that takes place during this period has implications for the teacher and coach and it can be considered the most important developmental period.

Puberty onwards

During this period great physical change takes place, with the young adult growing in size and increasing in strength. It must be remembered that there is great variation in the onset of puberty which can be as young as 9 years in some girls and as late as 15 years in some boys. When teaching and coaching young adults at this stage of their development, care has to be taken as their growth, strength, and skill level often progress at different stages. In terms of skill development it is during this stage that the coach or teacher can start to focus on tactics and more advanced skills as the young adult is able to cope at a higher level. Their understanding of movement will have increased and they will be more aware of the influence of the environment and task constraints on their movement.

Postural development

As already discussed in this section, the development of postural control is key to all other aspects of development. Postural control is necessary for stability, balance and orientation. Once a child can hold up its head they can begin to explore their environment, and once they have the postural strength to hold themselves in an upright position they can then sit and then begin to crawl and then walk. The development of posture can therefore be considered in three phases: head control, sitting and standing.

Reflexes (see Section M) such as the righting reflex (one of the postural reflexes) enable the child to maintain a stable head position from birth. However, newborns have very little control over their posture. More voluntary control occurs from approximately 8–10 weeks of age when the young infant can keep its head steady when perturbed (with the body supported). However, by approximately 3 months, more consistent postural responses are made by the head in response to perturbations (Prechtl, 1986). This indicates that voluntary head control is fully achieved as the child is able to make adjustments of the head in relation to changes in body position. This is not easy to achieve as the child's head is heavy and out of proportion to the rest of the body. It must also be remembered that control of the head is important as two sensory systems, the visual system and the vestibular system, are in the head. At 3 months of age most children can sit without support; however, at this age if the baby tilts to one side then it would struggle to regain control of its head. Sitting is generally achieved in two stages, with the child sitting when supported and then sitting unsupported. At approximately 5 months of age most children have the level of postural control needed for independent sitting. This is an important stage in the child's development as it enables them to be in a better position to explore the environment. Sitting involves both balance and postural control; in addition, not only does it involve joints muscles and coordination, it is also a perceptual achievement. Maintaining a sitting position involves balance, and the infant must also detect perceptual information and integrate it with appropriate motor responses. Stationary postures such as sitting may appear to be static, but the body is continually in motion. To maintain balance, a body sway in one direction by the infant must be followed by a compensatory sway response in the opposite direction. Infants who have just learnt to sit tend to fall over when their body moves outside the limits of the sway region, and they do not have sufficient strength and coordination make an appropriate correction. You may have observed an infant who has just learnt to sit, falling over; they tend to be very rigid as they fall. At this stage they cannot make compensatory movements or turn their heads or twist their body to counter the fall. *Figure 1* shows the stages of postural development that children typically go through.

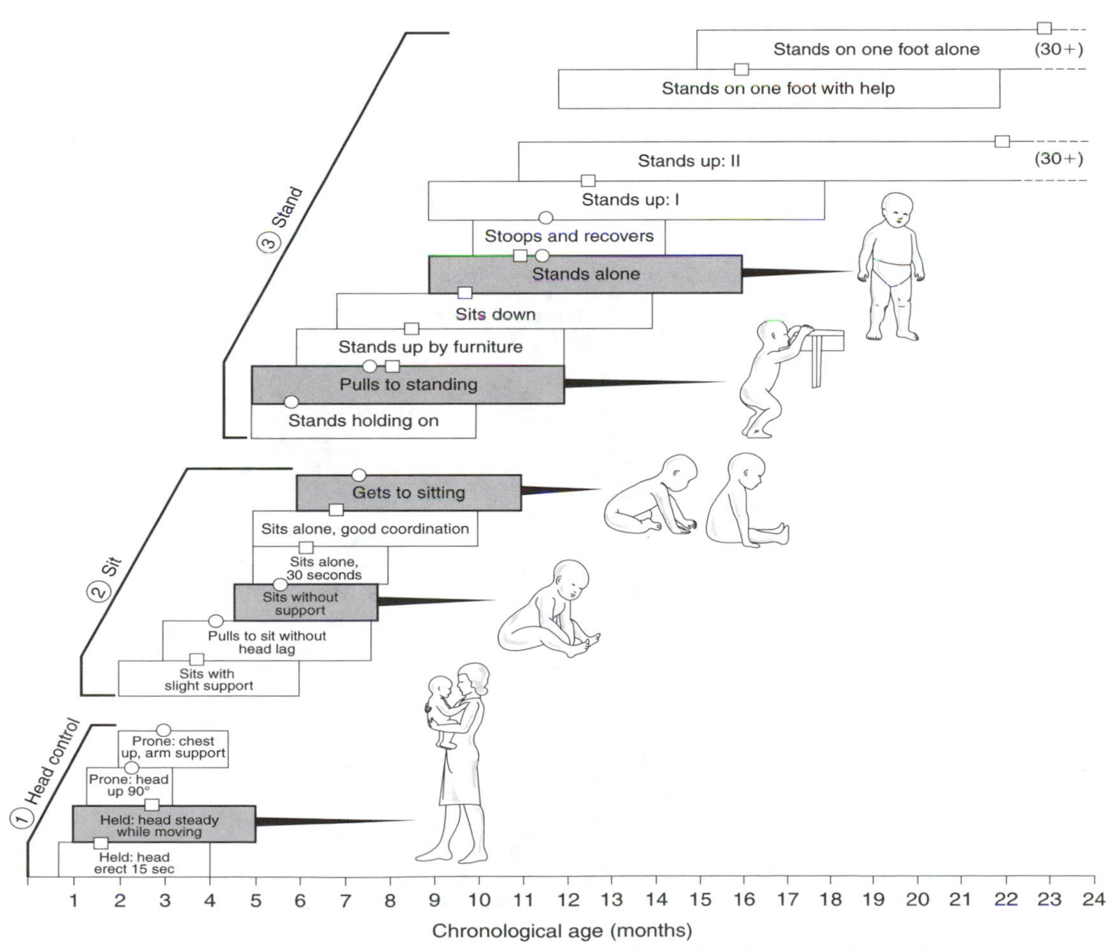

Fig. 1. The progression of postural control. Reproduced with permission from Keogh and Sugden (1985).

As the child develops they need to control their head and neck and be able to make adjustments if they move. Woollacott *et al.* (1987) noted that by the time infants are 8 months of age the muscles in the neck and trunk were coordinated into effective patterns for controlling forward and backward sway in the seated position as indicated by the formation of muscle synergies. With this emergence of head and trunk control, the infant now begins to be able to change body positions and from supine or prone, and can achieve sitting positions. These actions culminate with the infant being able to stand independently at approximately 11 months of age (Keogh and Sugden, 1985). Without the ability to control their heads, children would not be able to achieve this upright position. Prior to standing alone the infant often uses furniture and other people to pull themselves up to stand, utilizing environmental affordances to achieve their movement goal.

During the process of learning to stand independently (*Fig. 2*), infants must learn to balance in more demanding situations compared to those during sitting and they also have to control many additional degrees of freedom as they add the coordination of the leg and thigh segments to those of the trunk and head (Shumway-Cook and Woollacott, 2001). Standing is generally achieved in two phases: firstly, standing supported (cruising) and secondly, standing unsupported. Gaining postural control is a key aspect of achieving a standing position. Longitudinal studies by Woollacott and Sveistrup (1992) and Sveistrup and Woollacott (1996) showed that independent stance occurs due to the activation of directionally appropriate postural synergies. Infants at 2–6 months of age (prior to pull-to-stand behavior), did not show coordinated muscle response organization in response to threats to balance. However, as pull-to-stand

Fig. 2. During the first nine months most children will stand with some support or independently.

behavior progresses, the infants begin to show directionally appropriate responses in their ankle muscles. As the pull-to-stand skills improve, muscles in the thigh segment were added and a distal-to-proximal sequence began to emerge in infants 9–11 months of age, as independent stance eventually emerges. Chena *et al.* (2006) have found that in infants postural sway when sitting increases greatly prior to achieving standing. The authors indicate that this may well result 'from a process involving re-calibration of an internal model for the sensorimotor control of posture so as to accommodate the newly emerging bipedal behavior of independent walking'.

Researchers state that the emergence of adult levels of postural control occur at different times for different aspects of postural control. However, by 7 to 10 years of age a child's postural control is essentially adult-like. Many key changes contribute to this refinement of postural control occurring between the ages of 2 and 7 years. Studies show that as the child gets older and their balance improves, the amplitude of sway decreases (Hayes and Riach, 1989). In addition, refinement of compensatory balance adjustments in children improves with age, and well-organized muscle synergies are produced in response to postural perturbations (Shumway-Cook and Woollacott, 1985).

In children as in adults, coordination between posture and movement can be achieved in either in a feedback manner or a feedforward manner. During childhood, it is classically admitted that feedforward control matures later than feedback control (Bernstein, 1967; Kelso, 1982), but when feedforward control does emerge, adult-like postural control is observed. Developmental studies involving postural control suggest that feedforward control, despite its early emergence, takes time to develop and the most mature components do not emerge until the child is 4 to 5 years of age (Hay and Redon, 1999, 2001; Assaiante *et al.*, 2000). Moreover, the expression of anticipatory locomotor adjustments for obstacle avoidance is still maturing during mid-childhood (McFayden *et al.* 2001). Until the development of anticipatory postural control is completely mature, the stabilizing framework that supports the voluntary control is inefficient, thus control during performance skilled action deteriorates. The experimental work of Schmidtz *et al.* (2002) suggest that anticipatory postural adjustments are not fully developed until the age of 8 years, although improvement of the performance of the postural stabilization clearly appeared between 6 and 7 years of age.

From a dynamical systems approach, postural control emerges from a complex interaction of the musculoskeletal system and neural systems in relation to the individual, task and environment. The achievement of a level of postural control that enables the child to stand independently means that the child is able to cope with controlling multiple degrees of freedom in a situation that has less stability than sitting. A number of key developments take place that facilitate this change in postural control. The child develops the leg strength to hold themselves upright. Postural response synergies are developed that enable the child to make appropriate postural responses when they move forward or backwards. Sveistrup and Woollacott (1996) found that infants who had not yet been able to achieve a pull-to-stand position did not show coordinated muscle responses when perturbed. However, during early pull-to-stand behavior children did begin to show coordinated muscle responses when perturbed.

Once a child can stand independently and is able to make some adjustments when they lose balance, they are ready to walk. The next part of this section will consider the development of locomotion with the focus on walking and running.

Development of locomotion: crawling, cruising, walking, and running

The development of locomotion or mobility is a key motor landmark that is especially important for parents (*Fig. 3*). Some of the upper limb developments that a child makes are just as important but they are also more difficult to observe. The onset of walking, however, is very visible and rapid progress is made. Most importantly it enables the infant to make whole-body movements and to explore their environment in a whole new way. Mobility is a skill we take for granted, but a skill that we use in a variety of ways in our daily lives. We all walk, run, skip or wheel around on a daily basis and are able to do this in a wide range of contexts in order to complete a wide range of tasks. Not surprisingly, locomotion has been the focus of many studies from a range of academic fields. This section will consider a variety of locomotive skills starting with crawling.

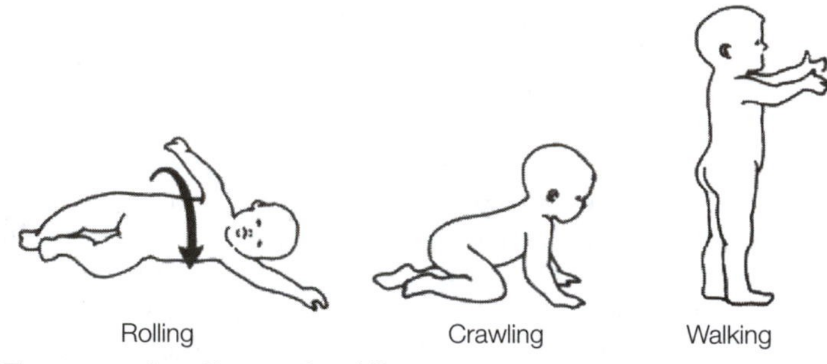

Fig. 3. The progression of locomotive ability.

Crawling

The first locomotive movement seen in infants is rolling, and this occurs between 4 and 10 months of age. However from approximately 6 months of age infants begin to move forward in a prone position by some form of creeping or crawling (*Fig. 4*). Crawling is very much a pre-walking progression which enables the child to start exploring the environment in a more dynamic manner. In developing a crawling action infants display a variety of techniques. Adolph *et al.* (1998) observed that some infants passed through a stage of belly crawling before crawling on hands and knees while others did not. In a range of studies (see Section M for more detail) looking at crawling in infants it has been found that infants have to identify new parameters for control when they start to crawl. Adolph (2000) found that infants new to crawling would crawl 'over the brink' on a real visual cliff task. However, the same infants who were experienced at sitting would not reach to a point that would make them overbalance. This action-perception coupling is discussed in more detail in Section M. The next stage in gaining mobility is termed cruising.

Cruising

The next stage on the path to full-scale walking and house destruction is termed cruising. Once the infant has the leg strength and the balance control, they will use objects such as chairs to pull themselves up and to support themselves. As their strength and ability to balance increase infants in a supported position take less weight on their arms and more on their legs (Haehl *et al.*, 2000). Eventually they will use only one hand when cruising. Cruising therefore is a progression that aids the development of balance and assists the infant in gaining independent walking.

Fig. 4. Young infants crawling.

Walking

At around 12 months of age, with great variation, infants will begin to walk independently. Malina (1980) reviewed numerous studies on onset of walking, and states that the median age was between 11.4 and 14.5 months. Early walking is very hesitant, with the child using a wide stance often with the feet turned out to assist balance. Steps tend to be flat-footed and short, with width of stance often longer than step length. The arms are held in a fixed high position which is referred to as the 'high guard'. This not only gives the child fewer degrees of freedom to control but also acts as a safety mechanism.

If the child falls the hands are ready to catch them and protect the head. (A detailed theoretical explanation of why children start to walk can be found in Section M). The refinement of walking takes place quite quickly. The arms are gradually lowered, followed by a flexed arm swing before the child adopts a reciprocal arm action. Leg movements become more continuous and the wide stance is reduced and more forward movements are made. This occurs after approximately 4 to 6 months of independent walking. Interestingly, infants are very sensitive to perceptual information. Lee and Aronson (1974) conducted experiments in which the walls of a room were moved in order to disrupt optical flow information. They found that when the walls were moved beginner walkers would fall when both the front and, especially, the side walls were perturbed. Older more experienced walkers would stagger only after the side walls were moved. These experiments clearly indicate that infants are able to differentiate optical flow information.

A number of researchers have examined walking, but the work of Shirley (1931) and McGraw (1940) deserves a special mention. Shirley (1931) followed 25 babies from birth to 2 years of age. Movement observations were made and recordings of walking patterns taken. Shirley found four stages of walking development based on footprint data, and standing with support was considered to be the most important achievement. McGraw (1940) examined the progression to independent walking and identified seven stages of erect locomotion from reflex stepping to an adult pattern of walking. As Shirley had done previously, McGraw made numerous observations on large numbers of infants. More recently researchers have examined walking in infants using more advanced technologies such as 3D analysis and electromyography (EMG). Thelen and coworkers (Thelen and Fisher, 1982, 1983; Thelen *et al.*, 1982, 1984) provided detailed descriptions of infants kicking and stepping actions. More recently Clark and Phillips (1993), van der Meer *et al.* (1995), Assaiante *et al.* (2000) and Adolph *et al.* (1993) have conducted a range of studies that add to our knowledge of walking behavior in infants. Many aspects of walking gait have been examined, including inter- and intra-limb coordination, acceleration profiles and velocity profiles. It is clear that increased strength is vital for walking development and that the infant has to be able to support themselves on one leg and have the power to push the other leg forward during single-leg stance (Adolph and Avolio (2000). Brenière and Bril (1993, 1998) found that the vertical acceleration of infants' center of mass is negative at foot contact. This value gradually increases during single-leg stance as postural control develops. However, it is not until 4 to 6 years of age that positive values similar to those of adults are found. Other studies have also found that 4 to 6 years of age is a key time for walking development. Sufficient anticipatory leg muscle activation occurs at this age (Assaiante *et al.*, 2000), bilateral coordination is more stable (Loovis and Butterfield, 2000) and children have stable but well-coordinated movements between the arms and legs (Getchell and Whitall, 2003). What is clear is that more research is still needed in this area; for a full review see Whitall (2003).

Young children continue to refine their walking up to the age of 7 or 8 years. However, the changes are very discrete and difficult to assess by direct observation. During this time legs and arms are alternated, feet are placed more in line, they walk faster and spend less time with one foot in contact with the ground (Keogh and Sugden, 1985). They are also better able to make walking variations such as walking backwards or on tiptoe. Young children are also able to cope with environment changes such as walking in the snow, climbing stairs, or walking downhill. Wickstrom (1977) indicated a number of important factors in walking development. These include:

- move towards heel strike rather than flat-footed strike
- toe push off
- decrease in hip flexion
- achievement of a double-locking knee.

Once the young child has reached a competent level of walking, the next stage is more dynamic movements such as running and jumping. You have to walk before you can run and the development of running will be considered next.

Running

By 2 years of age most children are able to run, in fact most children achieve this at approximately 18 months of age. Walking and running are very similar, but the main difference is that when running some time is spent with both feet off the ground. This presents an interesting coordination problem for the young child. By the ages of 4 to 6 years most children are able to run with good control and coordination.

In the initial stage of running (*Fig. 5*), young children tend to have a wide base of support, an incomplete extension of the supporting leg and a limited flight phase. Movements of the legs are limited with a short leg swing and uneven stride (Gallahue, 1989). In the recovery phase the leg rotates outward from the hip, and the swinging foot is often turned outwards. Movements of the arms tend to be short and stiff with varying degrees of elbow flexion. By approximately 2 to 3 years of age the young child will develop a more advanced running style that Gallahue (1989) refers to as elementary. The base of support will not be as wide and there is a more complete extension of the support leg at take-off and more knee flexion during recovery. There is an increase in length of stride, arm swing and running speed. By 4 to 6 years of age the child typically demonstrates a mature running style. Speed of running will have increased as will stride length. There will now be complete extension of the support leg and a definite flight phase. The recovery of the leg and foot will involve minimal rotation and the arms will swing vertically in opposition to the legs. Other factors that are of note are the change from a full sole contact to a heel–toe contact. The path of the center of gravity fluctuates less, becomes smoother and has a more forward action. Finally, less time is spent with the foot in contact with the ground, and more time in flight. Once this technique is acquired the child will continue to increase their running speed into late adolescence.

Fig. 5. Stages in running development. Adapted from Gallahue (1989).

It must be remembered that locomotion is complex, as is the interaction between the individual, the task and the environment. When moving around, children have to also avoid objects and perform tasks such as crossing roads. Before we consider manual skills, tasks of this nature need some consideration. Pryde *et al*. (1997) looked at locomotor ability and obstacle avoidance in children and adults. Movement time, number of errors and qualitative measures were taken on children's performance on three routes that varied in complexity. One was a straight pathway requiring walking only, the next a twisting pathway requiring steering ability, and a final pathway had obstacles. Children below the age of 11 years displayed significantly slower movement times when they were required to avoid obstacles; however, children under the age of 8 years made significantly more errors in the obstacle path and tended to take larger than necessary steps over obstacles. Before the age of 8 years, children's locomotive ability is therefore challenged by more complex environments. Other studies have considered children's ability to move in more complex environments by looking at road crossing. This is obviously an increasingly important aspect of children's daily lives, and we know that children are at great risk of being involved in road accidents. Te Velde *et al*. (2003) provide an overview of road-crossing behavior in young children. They state that children's ability to select appropriate places to cross the road is not developed until after the age of 9 years. More interestingly, they conclude that children's visual search strategy is not fully developed by the age of 7 to 8 years, and only by the age of 11 is it more like that of the adult. A number of changes therefore take place between the age of 7 and 11 years, with younger children showing more caution, and movement experience being a key factor, with great individual differences.

Development of manual skills: grasping, reaching and grasping, catching, and throwing

Manual control refers to an individual's ability to voluntarily control and complete a desired task or action involving reaching for, grasping and manipulation of an object. Manual skill is a general term for the arm and hand movements used to control objects. The arm and the hand are the primary units, with the arm having the functions of support, positioning, and force. The functions of the hand are grasp and manipulation (*Fig. 6*). The development of manual control is complex and involves the development of many behaviors. During the first year of life the manual system develops rapidly; the ability to manually control an object develops as the infant moves from a crude hold to a more controlled grasp. The infant's ability to reach and grasp an object accurately and consistently increases as the system of control develops and is refined. By the end of the first year infants can pick up objects and manipulate them (von Hofsten, 1986). During the second year the child is capable of using sensory information to make adjustments (Gordon, 1994). By 6 to 8 years of age the reach and grasp resembles that used by adults and continues to be refined during adolescence. Manual control and co-ordination form a fundamental part of our daily movements. Using manual control we reach, grasp, hold and manipulate various objects in a number of ways with varying levels of force, some simple, others complex.

Manual control can be unimanual (one arm/hand only) or bimanual (both arms/hands working together). When the hands are working bimanually they may be performing the same actions (symmetrical) or they may be employing different actions (asymmetrical). For example, one hand could be supporting an object while the other is manipulating part of the object. Guiard (1987) emphasized that in most skilled manual activities both hands are performing different

Fig. 6. Young children are able to use both hands to complete a range of manual tasks.

tasks at the same time. Over time young infants become very skilled at both unimanual and bimanual actions.

Grasping

During the first months of life interactions between the infant and the environment mainly involve the hands. The ability to grasp can be seen once the baby is born (even when premature); a baby can automatically take hold of an object placed in its palm. This reaction is known as the grasp reflex (Twitchell, 1970) as discussed earlier. During the first few weeks of life some infants use a crude hand posture that assists grasping, but most use a hand posture that does not assist grasping. This is achieved at around 5 to 6 months of age. The release of objects is unpredictable and influenced by the grasp reflex, resulting in poor voluntary release. However, by 8 months the child is able to release objects. Initially this is just an opening of the hands to release the object. At around 10 months some control of the release is gained. Grasping smaller objects requires the opposition of the thumb with one finger. This requires a pincer grip and can be achieved by children at around 9 to 10 months. This important stage in the development of manual control leads to the development of the precision grip; it is this development that allows the digital control of objects. Forssberg *et al.* (1995) found that young children tended to use a high grip force to grasp objects using a precision grip. This gave a high safety margin and was considered to be a reflection on the reduced ability to adapt grip force to the load force based on tactile information. During manual skill development there are phases through which the child passes. The first focuses on hand control, the second sees changes in the arm linkage system and the third is an extension of the first two, but here the emphasis is on self-help skills. *Figure 7* gives an overview of the

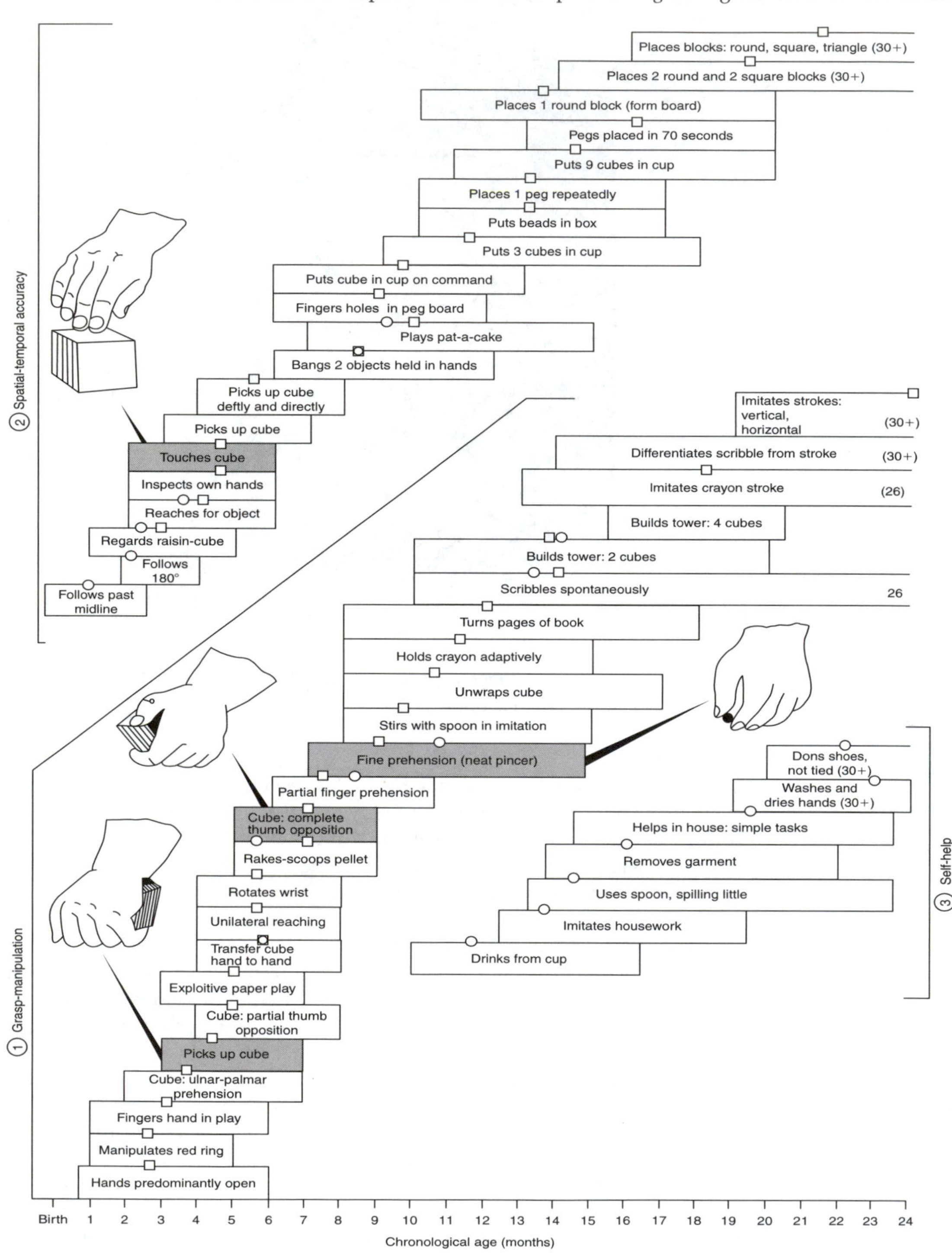

Fig. 7. Development of manual skill. Reproduced with permission from Keogh and Sugden (1985), p.46.

development of both grasping and reaching and grasping. Then during the next 6 months the child's manual control develops in terms of the manipulation of various objects. Finally, the manual control and skills that have developed are used in a wide range of activities that enable the child to feed, dress, and grasp more complex objects (*Fig. 8*).

Elliott and Connolly (1984) have categorized manual skills into three classes: simple synergies, reciprocal synergies and sequential patterns, each of which has a range of movement patterns within it. The categories of movements used are functionally based and include most manipulative movements. The first distinction made between movements is whether they are simultaneous or sequential. In simultaneous movements the digits are used together in a single movement pattern, in sequential movements there is an involvement of the digits in an independent manner. The first two categories of simple and reciprocal synergies both involve simultaneous movement patterns. Simple synergies (pinch, dynamic tripod and squeeze) are those that involve all the participating digits, including the thumb, being convergent flexor/extensor synergies, such as in squeezing a rubber ball. Reciprocal synergies (twiddle and rock), in comparison, involve movements in which the thumb and participating digits perform dissimilar movements. An example is the flexion of the fingers and extension of the thumb that occur in the fine manipulation of objects, such as rolling a piece of plasticine between the thumb and first two fingers. The third category, sequential patterns (rotary step, digital step, and linear step) involves a discontinuous movement of the object with some digits supporting the object while

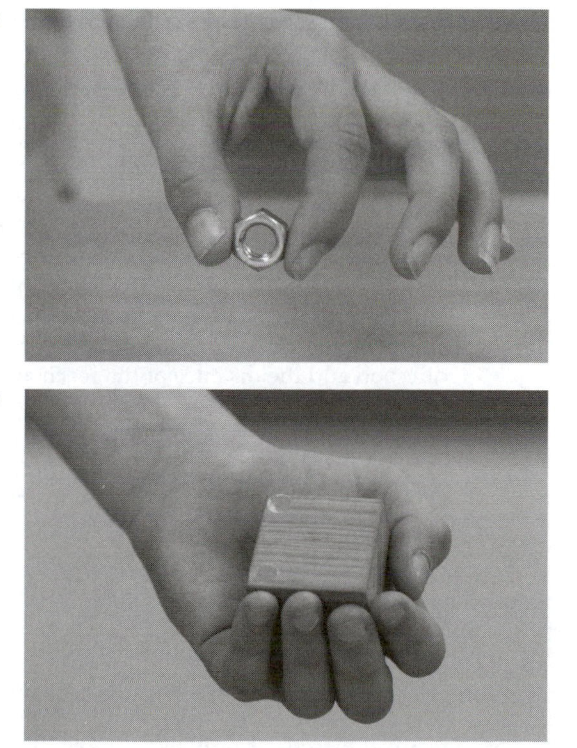

Fig. 8. Precision grip (a) and palmar grip (b).

others reposition it, which in turn become the supporters while the former are involved in repositioning the object. Turning a pencil around in the fingers is an example. By describing manual skills in this way, movements can be both described and assessed in detail. In order to engage in a wide range of activities, not only do we have to grasp but we also have to reach (*Fig. 9*). The next part of this section will therefore deal with reaching and grasping.

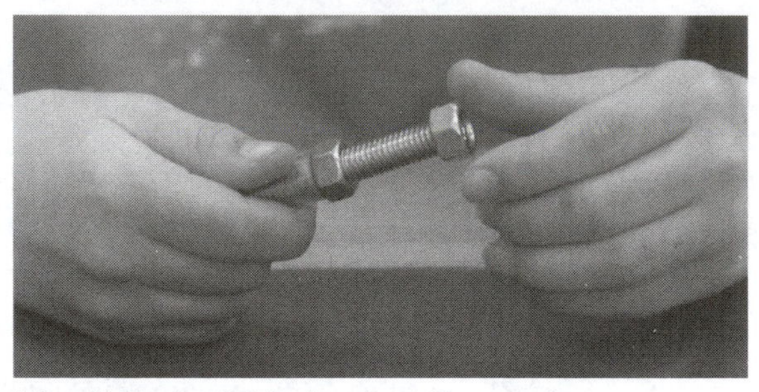

Fig. 9. Many daily activities involve the hands working together.

Reaching and grasping

By observing the reaching and grasping of infants the development of many aspects of manual control can be seen. Keogh and Sugden (1985) considered reaching and grasping as three general achievements (p.48):

1. differentiating reach and grasp into re-combinable movements
2. improving reach in spatial accuracy and efficiency
3. modifying grasp from palmar to digital.

White *et al.* (1964) described the development of reaching in stages. They stated that the newborn child has an innate eye–head component which allows the infant to localize and track objects with head and eyes. Gordon (1994) stated that the newborn infant can initially make partially directed movements towards objects but these are very basic movements. What is clear is that young infants (under 6 months) can make crude reaches and grasps at objects.

Reaching and grasping are initially a single movement pattern with the reach and the grasp occurring at the same time (Bower *et al.*, 1970). There are two developmental milestones in the development of reaching and grasping according to Bower *et al.* (1970). The first is the ability to reliably preprogram the direction and the distance of the reach and grasp. The second, which is reached at around 9 months of age, is to reliably preprogram the orientation and the size of the object (Rosenbaum, 1991). At a very early stage (6–11 days) an infant can preprogram a reaching action in the correct general direction of an object. Bower (1972) reported that children as young as 7 to 15 days of age would not reach for objects that were out of reach. This indicates that they have some understanding or knowledge of distance at a very young age (*Fig. 10*).

The ability of human infants to reach for and retrieve objects in their environment develops slowly over the first 2 years of life (Berthier *et al.*, 1999). Infants progress from a crude ability to direct hand movements towards targets at about 8 weeks (von Hoftsen, 1982) to successful reaching at about 4 months of age. At first whenever the infant extends the arm the hand opens in extension simultaneously; however, at 2 months of age, this extension synergy is broken and the fingers now flex as the arm extends (von Hoftsen, 1984, 1993). Early on, objects are

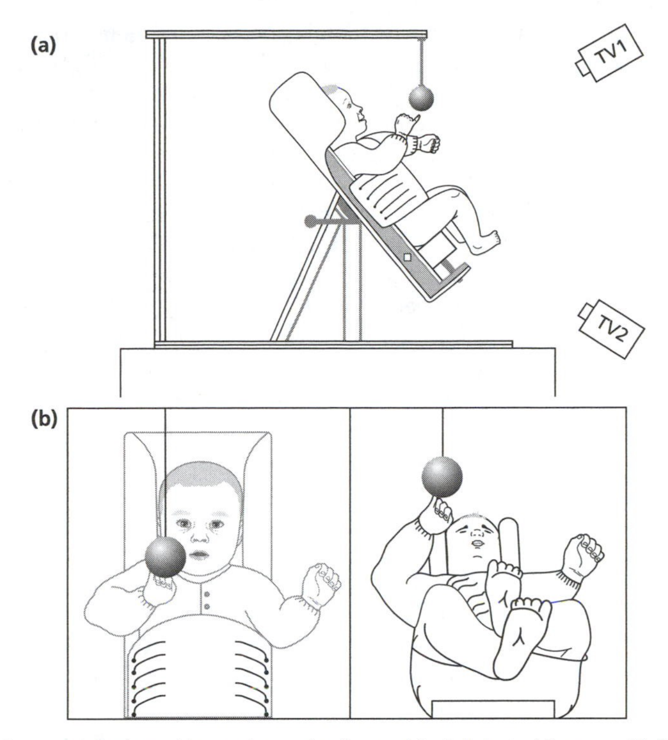

Fig. 10. A young infant reaching and grasping for an object. Adapted from von Hofsten (1982) by Shumway-Cook and Woollacott (1995), p.380.

grasped in a power grip, using the palm and palmar surface of the fingers, with the thumb reinforcing this grip (Bower, 1974; von Hoftsen, 1984; Forssberg *et al.*, 1991). At 9–10 months of age, pincer grasps are used to pick up small objects; there is complete thumb opposition and a precision grip is achieved allowing digital control of objects to be accomplished. Preceding this achievement, Knobloch and Pasamanick (1974) note that there is a crude release of the object occurring at around 8 months of age. Further development of manual dexterity allows this release to be under more control and objects can now be placed.

Konczak and colleagues (Konczak *et al.*, 1995, 1997) studied infants from 4 to 15 months of age during reaching actions and showed that there are two developmental phases in hand trajectory formation. The first developmental phase occurs at 4 to 6 months of age; it must be noted that at this time infants gain more trunk stability and strength in the neck muscles, thus although the reach becomes more accurate it is still a little segmented (Schneiberg *et al.*, 2002). In the second developmental stage, there is fine tuning of the sensorimotor system, in which there are more gradual changes in endpoint kinematics; it is noted, however, that at 3 years of age endpoint kinematics have still not reached adult level of movement.

By the age of 2 years, infants have well-coordinated movement sequences in the arm–hand linkage and they are more spatially accurate in placing the hand where desired. The majority of research in reaching ability has been done in children under the age of 3 years (Schneiberg *et al.*, 2002). These studies mainly focus upon analysis of movement time, movement segmentation, hand trajectories temporal aspects of interjoint coordination and joint torque (Thelen and Smith, 1994; Konczak *et al.*, 1995, 1997; Savelsbergh *et al.*, 1997). Able-bodied children acquire the ability to co-regulate trunk and arm movements for functional activities over the first 10 years of life, and evidence suggests a developmental transition period occurs between the ages of 4 and 7 years. A number of studies have shown that between 4 and 6 years of age children can make movements without visual feedback with reasonable accuracy. However, at 7 years, there is an abrupt reduction in this ability, reflecting an increased dependence on visual feedback (Hay, 1979, 1990; van Dellen and Kalverboer, 1984).

Schneiberg *et al.* (2002) investigated the development of reach-to-grasp movements in children aged 4–11 years, and compared the kinematics of these with those of adults. Hand trajectories were noted to become smoother with age and less variable; children aged 7 years and younger had more curved trajectories than older children and adults, the further away the target was placed. By the age of 4 years, children use the same proportion of elbow extension as adults for reaching to close and far targets. Results also showed that interjoint coordination becomes more consistent, with mature patterns emerging at 8 years of age. However, only children between the ages of 8 and 10 years had variability at levels similar to those of adults. The high variability of interjoint coordination in younger children may reflect the system's attempts to search for optimal kinematic solutions (Thelen and Smith, 1994; Schneiberg *et al.*, 2002).

Multijoint goal-directed movements such as reaching and grasping can be performed in a variety of ways, thus it is remarkable that unrestrained prehension movements of adults show stereotypical kinematic features (Kalaska and Crammond, 1992). When we plan to pick up an object in the environment, our first goal is to contact that object at points along its surface that allow us to generate a stable grasp. Jeannerod (1981, 1984) was the first to provide a detailed description of the two components of human prehension. The accomplishment of this goal involves (1) spatial positioning of the arm – the transport component of the movement and (2) anticipatory shaping of the hand or the grasp component of the movement. On the basis of his early observations, Jeannerod (1981, 1984) argued that intrinsic object properties influence the shaping of the hands, and extrinsic properties influence the transport aspect of the movement. Jeannerod (1984) compared the reaching and grasping for objects of differing sizes and shapes. He found that size had no influence on the transport aspect of the reach and that maximum aperture increased as object size increased.

Performance of a standard prehension task has been found to show a characteristic pattern of hand transport and grasp aperture shaping, with the following general characteristics (Tresilian and Stelmach, 1997):

1. the tangential speed profile of the transport movement has an asymmetric 'bell shape', in which the deceleration phase is slightly longer than the acceleration phase; as accuracy demands increase the deceleration phase becomes comparatively longer (e.g. Martenuik *et al.*, 1990)
2. transport movement time (MT) increases with increasing target distance and target accuracy constraints that can be described by a Fitts' law-type relationship (Bootsma *et al.*, 1994)
3. the maximum grasp aperture increases monotonically to a maximum that is somewhat larger than is needed to grip the target (Martenuik *et al.*, 1990) and that is scaled according to the relationship between object size and hand size (van der Kamp *et al.*, 1998)

4. errors of hand transport that may occur during prehension without visual guidance are compensated by a widening of the grip aperture (Wing *et al.*, 1986)
5. maximum aperture is always reached after maximum transport speed at between 60% and 80% of total precontact time, and is achieved at roughly the same time as peak deceleration of the transport phase of the limb (Zackowski *et al.*, 2002).

Although the general characteristics above describe hand transport and grasp aperture in normally developing children, many of these have been shown to be altered in children or adults with motor impairments. Maximum grasp aperture is usually larger, movement time longer and they exhibit much variability in their coupling between the grasp and transport phase (Smyth *et al.*, 2001; Zackowski *et al.*, 2002; Utley *et al.*, 2004).

Numerous experiments have been conducted in order to discover if the components of reaching and grasping are controlled independently or if there is a dependency between the transport and grasp phase. Smooth coordination of reaching and grasping is not innate but evolves gradually through ontogeny (Jeannerod, 1986). However, compared to the numerous analyses of prehension in human adults, few kinematic data are available concerning children. Kuhtz-Buschbeck *et al.* (1998) undertook a cross-sectional study producing a detailed kinematic analysis of reaching and grasping in healthy 4–12-year-old children, specifically addressing the development, coordination and refinement of these components.

Kuhtz-Buschbeck *et al.* (1998) showed that the refinement of hand transport occurred throughout the age range studied. This was evident from the emanation of a smooth and reproducible, bell-shaped velocity profile with a decreasing number of movement units per reach and more linear trajectories. Thus, although improvements were far less than those noted during the first 3 years of life, they may nevertheless represent the final stage of this development. Other kinematic features remained stable, showing no further development between the ages of 3 and 12 years for the task studied. Movement time was constant when vision was used, and appropriate scaling of movement velocity to movement amplitude was evident in all children. Anticipatory preshaping of the grip aperture according to target size was evident in the 4-year-olds, although the younger children's maximum aperture was relatively wider than the older children's.

Corresponding results have been found in infants (von Hofsten and Rönnqvist, 1988) and it is thought to compensate for inaccuracies of hand transport. When vision of the target was removed, preshaping of the grip was less consistent and adequate adjustments of finger aperture to object diameter, as known in adult grasping (Jakobson and Goodale, 1991), were only evident in the 12-year-olds. Between 4 and 12 years, grip formation improved with a uniform opening–closing sequence, with a single peak that was well synchronized with the deceleration of hand transport. Grip formation and hand transport formed a reproducibly coupled synergy in the older children, as if defined by a stable central coordinative organization (Bernstein, 1967; Ulloa and Bullock, 2003; Mason *et al.*, 2004).

As children get older they become more proficient in feeding and dressing, writing and drawing, and there is an increase in speed between the ages of 6 and 12 (Keogh and Sugden, 1985). By the age of 7 years they use a variety of grips and demonstrate the ability to produce various combinations of arm and hand movements. These movements become more continuous, appear to be easier and are smoother. A simple reach-and-grasp movement is well within the capability of most children. Gordon (1994) stated that between the ages of 6 and 8 years, children demonstrate a level of control that is similar to that of an adult. Forssberg *et al.* (1991) conducted studies examining grip force and object weight during reaching and grasping. Children between the ages of 10 months and 12 years were compared to adults. It was found (using a variety of measures such as peak load force and vertical acceleration) that by 6 to 8 years of age the results started to match those produced by adults.

Catching

Catching has been defined as the action of bringing an airborne object under control using the hands and arms (Payne, 2001) or a coincident timing task that involves the complex interplay of coordinating visual information and motor behavior to a single point of interception (Gabbard, 2002). Catching behavior has been demonstrated in young infants of 16–20 weeks of age. Von Hofsten and colleagues (von Hofsten and Lindhagen, 1979; von Hofsten, 1980, 1982, 1984) provide empirical evidence of the ability of infants to catch moving objects. Children were placed in a semi-reclining infant seat and an attractive, colorful object traveled in front of them at chest level. The speed and distance the object moved was manipulated and the age of the participants varied from 4 days to 9 months. The von Hofsten studies showed that infants are able to anticipate the future location of a moving object and that infants used a strategy that consisted of an approach component and a tracking component (*Fig. 10*). In the von Hofsten and Lindhagen (1979) study, the infants seemed to display their catching sophistication by not automatically reaching toward the object independent of its traveling speed. The infants knew whether or not a catch by themselves was possible or not in that when the objects approached at a high velocity a catching attempt was not made.

Such results are in line with the results provided by Savelsbergh *et al.* (1989); this study investigated whether the positioning of an object would affect which hand a child would use to attempt to catch an object. Three infants (2.5, 5 and 10 months) sat in a semi-reclining seat and a colorful ball was presented to each child in two conditions. In the static condition, the ball is placed on a table in front of the child within reaching distance at three different positions, thus requiring reaching behavior. The second condition, the dynamic condition, required catching behavior of the infant. The ball was attached to a pendulum and was made to approach the infant at chest level from the three different positions used in the static condition. In contrast to the von Hofsten studies, the ball approached the infant in the frontal plane, the time window therefore being more limited. A clear difference as to whether or not an attempt was made was noticed between the two youngest infants in the static condition and between the two youngest and the oldest infant in the dynamic condition. The infant aged 2.5 months showed only reaching behavior but no catching behavior. The infant at 5 months showed catching behavior but the frequency was less than that of the 10-month-old and was also more dependent upon the position the ball came from. The 10-month-old was very able at catching the ball. Thus, despite the very limited sample size these results seem to support the contention of von Hofsten and Lindhagen (1979) that infants seem to be able to perceive affordances of moving objects.

The skill of catching develops gradually throughout childhood (Savelsbergh *et al.*, 2003). A number of researchers have outlined the emergence of catching behaviors across time, concluding that as children mature, their ability to proficiently catch a ball improves (Wellman, 1937; Kay, 1970; Strohmeyer *et al.*, 1991; Williams, 1992). Kay (1970) found that a 2-year-old child will hold their hands together palms up, and their eyes will watch the thrower or the thrower's hands, the ball falling into arms with no closure of the hands. A marked change in strategy was noticed in a child aged 5 years. The 5 year old watched the ball for the entire flight, and then moved the hands to meet the ball with the hands. By contrast, a 15-year-old moved to intercept the ball and grasped with the fingers. A clear developmental change was that visual attention at the youngest age centered mainly on the origin of the ball's motion. At 5 years of age, attention was focused upon both the ball and the catcher's own hands, and by 12 years of age the attention had switched again to predominantly the ball's flight (*Fig. 11*).

Williams (1992) extended the work of Kay (1970), carefully recording and describing the adaptations in the way children between the ages of 4 and 10 years used their visual attention and their limbs to catch a ball that was thrown to them by another person. Three distinct visual strategies were observed across the age range. The strategies were described as retrospective (least mature), concurrent and predictive (most mature). Also, three distinct movement strategies were noted, namely cradling (least mature), clamping and grasping (most mature). The strategies used by the participants were coded; each category of

Fig. 11. As children develop, their movements are less rigid and more adaptable, indicating a release in degrees of freedom.

visual and movement strategy was considered as a developmental 'coupling'. Williams concluded that a reasonable expectation might be that the visual and motor strategies are coupled at the appropriate level of maturity. Thus, the movement and visual strategies were coupled as: retrospective–cradling, concurrent–clamping and predictive–grasping.

Exclusive coupling was only witnessed in 10-year-olds when catching with both hands, a predictive–grasping visuomotor strategy much like that seen in the adult was used at all times. At all other ages (4–9 years) visual and movement strategies were varied; this was particularly evident in the 6–8-year-olds. It seemed these children were experimenting with the various ways and means they had at their disposal to prevent the ball being missed. Interestingly, if the ball was dropped, a child would often adopt a less-mature movement pattern, often a cradling action, as this represents the most economical efficient strategy as body parts are used as a functional whole. Utley and Astill (2007) conducted a study examining two-handed catching in children aged 7 to 8 years with developmental coordination disorder (DCD) and their age-matched controls (AMCs). Kinematic data were collected to examine Bernstein's (1967) notion of freezing and releasing degrees of freedom. Kinematic analyses showed that children with DCD exhibited smaller ranges of motion and less variable angular excursions of the elbow joints than their AMCs, and that their elbows were more rigidly coupled. These data suggest that children with DCD rigidly fix and couple their limbs to reduce the number of degrees of freedom actively involved in the task.

Other published research has noted the movement strategies noted by Williams (1992). Haubenstricker *et al.* (1983) and Strohmeyer *et al.* (1991) noted that first attempts to catch an aerial ball are typified by a more passive approach; the child will hold his/her hands rigidly extended forward and the ball is usually trapped against the chest. As the child becomes more proficient at catching there is extension of the arms forwards and sideways, either hugging or scooping the ball to safety. The child's limbs have become more active participants in the attempts of the child to catch the ball. In addition, there is awkward adjustment of the body as the catcher 'fights' to remain balanced. A more proficient catcher responds with a series of coordinated movements involving highly precise forms of spatial adjustment and coordinative action (Kay, 1970; Haubenstricker *et al.*, 1983, Strohmeyer *et al.*, 1991). Although many aspects of mature catching behavior are related to the specific catching task, Wickstrom (1983) indicated that in general moving the hands into an effective position to receive the ball, and grasping and controlling the ball with the hands are two characteristics of a mature catcher. As the child begins to understand how to interpret meaningful information from the environment they will adjust their whole body to the flight characteristics of the ball. The feet, trunk and arms will all move to adjust to the path of the oncoming ball, the palms will be adjusted to accommodate the size of the ball and the height of the flight path, and the arms will fully extend to meet the ball.

As the work of von Hofsten (1982) shows, catching ability starts very early in life. However, Alderson *et al.* (1974) observed considerable changes in catching ability between the ages of 7 and 12 years. Alderson *et al.* (1974) noted that the ability to locate the hand in the correct position seems to be an important factor in determining catching performance. The catching efficiency of children aged 7 years, in comparison to those aged 10 and 13 years is believed to be due to the late initiation of flexion when attempting to catch one-handed. Alderson *et al.* (1974) also noted children aged 7 years found a one-handed catch too difficult and would have preferred to catch with two hands, while those that did catch with the one hand exhibited problems in the temporal prediction of grasping action.

Fischman *et al.* (1992) support the work of Alderson *et al.* (1974), as they found that catching performance improved with age. By the age of 5 years, the rudiments of skill in one-handed catching had emerged; by the age of 12 years, however, children had essentially mastered this skill. Fischman *et al.* (1992) noted that ball location appears to be an important task constraint in catching performance. The approach of the ball in their study was either above the head, out to the side, or at shoulder height. The study showed that tosses that were farthest from the trunk region were much more definitive in forcing the appropriate use of hand orientation than those that were closer to the body. Thus, the location of the ball approach constrains the child's selection of an appropriate hand–arm orientation. Fischman *et al.* (1992) also found that some young children had appropriate hand orientation but were still unable to catch the ball. In contrast, older children caught the ball successfully even with an inappropriate hand orientation. Specifically, a combined total of 34 inappropriate hand orientations was observed for 11–12-year-old boys when the ball was tossed to the waist, yet in 97% of cases a successful catch was observed. These findings support those of Strohmeyer *et al.* (1991), who suggest that older children may have 'played' with the way they responded when the ball was tossed to less-challenging locations. As children become more skilful, they have more freedom to experiment with their responses. It is argued (Savelsbergh *et al.*, 2003) that these findings indicate a freeing of degrees of freedom as children explore the possible solutions available to them for the task.

One of the observable characteristics of a successful performance is that the movement of all of the components of the motor apparatus are controlled. Bernstein (1967) recognized that separate regulation of each of these components is impossible due to the non-linear nature of the interaction between them. He inferred that to be able to control all these components, or degrees of freedom, these movements all have to be coordinated. Bernstein contended therefore that coordination was the process of mastering the redundant degrees of freedom into a controllable system (Bernstein, 1967, p.127). When a task is first attempted, Bernstein stated that the novice will 'freeze' their degrees of freedom; that is reducing any angular excursions to a minimum. Additionally, task-specific linkages or coordinative structures (Turvey, 1977) may be made both between and within limbs in an attempt to solve the control problem. With practice, the novice then releases some of the degrees of freedom and the task-specific devices will be reconfigured, so the novice then has more freedom to experiment with their responses. The catching findings of Fischman *et al.* (1992) and Strohmeyer *et al.* (1991) indicate that as children become more skilful, they experiment more freely with their movements in an attempt to achieve the task goal. In Bernstein's terms this exploration can be described as a shift from freezing to freeing the degrees of freedom.

Ricken *et al.* (2004) investigated how children (aged 8 years) coordinated degrees of freedom when intercepting stationary and moving balls. When there was movement between the participant and the object, other than that caused by the reach itself, movement time and deceleration time of the reach component were extended. Ricken *et al.* (2004) showed that when reaching to intercept a moving ball the participants showed a lower peak velocity prior to contact. These results implied that children were sensitive to the potential change in impact resulting from participant or object motion, and modified their response accordingly. It was also reasoned that the contribution of additional degrees of freedom, such as the trunk, to reaching is influenced by the desired contact between hand and object. Ricken *et al.* (2004) showed that, like adults, children recruit additional degrees of freedom of the trunk, which are coordinated with the hand to produce movement that preserves an appropriate level of impact at hand/object collision.

Throwing

Throwing is a skill that infants have from an early age; once they are able to release they are able to throw. However, early throws lack control and are generally infants casting away objects they no longer wish to hold. Throwing involves the projection of an object by a variety of techniques. Throwing can be overarm, underarm and overhead with one or both hands used to throw a variety of objects. The skill of throwing is generally employed to send an object as far as possible or as accurately as possible. The development of throwing involves the gradual release of degrees of freedom as the infant begins by learning to control the arm movement by limiting the degrees of freedom in the trunk and lower body. Early throwing therefore involves rigid movement of the lower body and limited movement of the arm. However, in order to gain control or increase the distance of the throw, degrees of freedom have to be released. As can be seen in *Fig. 12*, Gallahue (1989) refers to early throwing as the initial stage, and it is limited to a small amount of movement in the throwing arm.

Fig. 12. Stages in throwing development. Adapted from Gallahue (1989).

The characteristics of throwing as described by Gallahue (1989) during the initial phase include a throwing action that is mainly from elbow with the elbow remaining in front of the body. The throw is little more than a push and there is little rotation of the body with the feet stationary and close together. As discussed earlier, this gives the child less to control. The next stage according to Gallahue (1989) is the elementary phase and as *Fig. 12* indicates a greater range of movement takes place. The elbow is flexed to a greater extent and the movement starts from behind the head. A wide stance enables more rotation at the hips and a shifting of the body weight. This is followed by the mature phase which is more like that demonstrated by adults, with a releasing and adjustment of the degrees of freedom of movement taking place. The arm is swung backwards in preparation with the opposite elbow raised for balance. The whole action involves a greater range of motion and controlled rotation. This is achieved between the ages of 7 and 9 years in most children. As throwing is a complex skill, it is appropriate that it should be subdivided into its component parts and not surprising that children develop the skill on a continuum.

Conclusion

This section has considered the development of a number of fundamental movement skills that we use in our daily lives. What is clear is that from birth to around the age of 9 years a massive amount of change takes place. This progression of change varies from individual to individual, and over time our ability to move and explore the environment is refined. By adolescence most of us are highly skilled in terms of movement ability and able to deal with complex and changing movement conditions. We have concluded with two sections on development as we feel that the complex topics addressed are best discussed once the reader has come to terms with the fundamentals of control of learning.

Further reading

Button, C. (2002) Auditory information and the coordination of one-handed catching. In: Davids, K., Savelsbergh, G., Bennett, S.J. and Van der Kamp, J. (eds) *Interceptive Actions in Sport: information and movement.* Taylor and Francis, London, pp.184–194.

Loovis, E.M. and Butterfield, S.A. (2003) Relationship of hand length to catching performance by children in kindergarten to grade 2. *Percept Mot Skills* **96**, 1194–1196.

Piek, J.P. (2006) *Infant Motor Development*. Human Kinetics, Champaign, IL.

Savelsbergh, G.J.P., Davids, K., Van der Kamp, J. and Bennett S. (2003) *Development of Movement Coordination in Children: applications in the field of ergonomics, health sciences and sport.* Routledge, New York.

van Hof, P., van der Kamp, J. and Savelsbergh, G.J.P. (2006) Three- to eight-month-old infants' catching under monocular and binocular vision. *Hum Mov Sci* **25**, 18–36.

References

Adolph, K.E. (2000) Specificity of learning: why infants fall over a veritable cliff. *Psychol Sci* **11**, 290–295.

Adolph, K.E. and Avolio, A.M. (2000) Walking infants adapt locomotion to changing body dimensions. *J Exp Psychol Hum Percept Perform* **26**, 1148–1166.

Adolph, K.E., Eppler, M.A. and Gibson, E.J. (1993) Crawling versus walking infants' perception of affordances for locomotion over sloping surfaces. *Child Dev* **64**, 1158–1174.

Adolph, K.E., Vereijken, B. and Denny, M.A. (1998) Learning to crawl. *Child Dev* **69**, 1299–1330.

Adolph K.E., Vereijken B. and Shrout P.E. (2003) What changes in infant walking and why. *Child Dev* **74**, 475–497.
Alderson, G.J.K., Sully, D.J. and Sully, H.G. (1974) An operational analysis of a one-handed catching task using high speed photography. *J Mot Behav* **6**, 217–226.
Assaiante, C., Woollacott, M. and Amblard, B. (2000) Development of postural adjustment during gait initiation: kinematic and EMG analysis. *J Mot Behav* **32**, 211–226.
Bernstein, N. (1967) *Coordination and Regulation of Movements*. Pergamon Press, New York.
Berthier, N., Clifton, R., McCall, D.D. and Robin, D. (1999) Proximodistal structure of early reaching in human infants. *Brain Res Exp Brain Res* **127**, 259–269.
Brenière, Y. and Bril, B. (1993) Posture and independent locomotion in early childhood: learning to walk or learning dynamic postural control? In: Savelsbergh, G.J.P. (ed.) *The Development of Coordination in Infancy*. Amsterdam, Elsevier, pp.337–358.
Brenière, Y. and Bril, B. (1998) Development of posture control of gravity forces in children during the first 5 years of walking. *Brain Res Exp Brain Res* **121**, 255–262.
Bootsma, R.J., Marteniuk, R.G., Mackenzie, C.L. and Zaal, F.T. (1994) The speed accuracy trade off in manual prehension: effects of movement amplitude, object size, and object width on kinematic characteristics. *Brain Res Exp Brain Res* **98**, 535–541.
Bower, T.G.R. (1972) Object perception in infants. *Perception* **1**, 15–30.
Bower, T.G.R. (1974) *Development in Infancy*. Freeman, San Francisco.
Bower, T.G.R., Broughton, J.M. and Moore, M.K. (1970) Demonstration of intention in the reaching behavior of neonate humans. *Nature* **228**, 679–681.
Chena, L.C., Metcalfe J.S., Jeka J.J. and Clark J.E. (2006) Two steps forward and one back: learning to walk affects infants' sitting posture. *Infant Behav Dev* **1**, 16–25.
Clark, J.E. and Phillips, S.J. (1993) A longitudinal study of the intralimb coordination in the first year of independent walking. *Child Dev* **64**, 1143–1157.
Elliott, J.M. and Connolly, K.J. (1984) A classification of manipulative hand movements. *Dev Med Child Neurol* **26**, 283–296.
Fischman, M.G., Moore, J. and Steele, K. (1992) Children's one-handed catching as a function of age, gender and ball location. *Res Q Exerc Sport* **63**, 349–355.
Forssberg, H., Eliasson, A.C., Kinoshita, H., Johansson, R.S. and Westling, G. (1991) Development of human precision grip 1: basic coordination of force. *Brain Res Exp Brain Res* **85**, 451–457.
Forssberg, H., Eliasson, A.C, Kinoshita, H., Westling, G. and Johansson, R.S. (1995) Development of human precision grip IV. Tactile adaptation of isometric finger forces to the frictional condition. *Exp Brain Res* **104**, 323–330.
Gabbard, C. (2002) Attentional effects and limb selection: implications defining handedness. *J Sport Exerc Psychol* **24**(S5).
Gallahue, D.L. (1989) *Understanding Motor Development: infants, children, adolescent*, 2nd Edn. John Wiley and Sons, New York.
Getchell, N. and Whitall, J. (2003) How do children coordinate simultaneous upper and lower extremity tasks? The development of dual motor task coordination. *J Exp Psychol* **85**, 120–140.
Gordon, A.M. (1994) Development of the reach to grasp movement. In: Bennett, K.M.B. and Castiello, U. (eds) *Insights into the Reach to Grasp Movement*. Elsevier, Amsterdam, pp.37–59.

Guiard, Y. (1987) Asymmetrical division of labor in human skilled bimanual action: the kinematic chain as a model. *J Mot Behav* **19**, 486–517.

Haehl, V., Vardaxis, V. and Ulrich, B. (2000) Learning to cruise: Bernstein's theory applied to skill acquisition during infancy. *Hum Mov Sci* **19**, 685–715.

Haubenstricker, J.L., Branta, C.F. and Seedfelt, V.D. (1983) *Standards of Performance for Throwing and Catching.* Paper presented at the Annual Conference of the North American Society for Psychology and Sport and Physical Activity, Asilomar, CA.

Hay, L. (1979) Spatial-temporal analysis of movement in children: motor program versus feedback in the development of reaching. *J Mot Behav* **11**, 189–200.

Hay, L. (1990) Developmental changes in eye-hand coordination behaviors: preprogramming versus feedback control. In: Bard, C., Fleury, M. and Hay, L. (eds) *Development of Eye Hand Coordination Across the Lifespan*. South Carolina Press, Columbia, pp.217–244.

Hay, L. and Redon, C. (1999) Feedforward versus feedback control in children and adults subjected to postural disturbance. *Brain Res Exp Brain Res* **125**, 153–162.

Hay, L. and Redon, C. (2001) Development of postural adaptation to arm raising. *Brain Res Exp Brain Res* **139**, 224–232.

Hayes, K.C. and Riach, C.L. (1989) Preparatory postural adjustments and postural sway in young children. In: Woollacott, M.H. and Shumway-Cook, A. (eds) *Development of Posture and Gait across the Lifespan*. University of South Carolina, Columbia, pp.97–127.

Jakobson, L.S. and Goodale, M.A. (1991) Factors affecting higher-order movement planning: a kinematic analysis of human prehension. *Brain Res Exp Brain Res* **86**, 119–208.

Jeannerod, M. (1981) Intersegmental coordination during reaching at natural objects. In: Long, J. and Baddeley, A. (eds) *Attention and Performance*. Erlbaum, Hillsdale, NJ, pp.153–169.

Jeannerod, M. (1984) The timing of natural prehension movements. *J Mot Behav* **16**, 235–254.

Jeannerod, M. (1986) Mechanisms of visuomotor coordination: a study in normal and brain damaged children. *Neuropsychologia* **24**, 41–78.

Kalaska, J.F. and Crammond, D.J. (1992) Cerebral cortical mechanisms of reaching movements. *Science* **255**, 1517–1523.

Kay, H. (1970) Analysing motor skill performance. In: Connolly, K. (ed) *Mechanisms of Motor Skill Development*. Academic Press, New York, pp.152–164.

Kelso, J.A.S. (1982) *Human Motor Behavior: an introduction*. Erlbaum, Hillsdale, NJ.

Keogh, J.F. and Sugden, D.A. (1985) *Movement Skill Development*. Macmillan, New York.

Knobloch, H. and Pasamanick, B. (1974) *Gesell and Armatruda's Developmental Diagnosis*, 3rd Edn. Harper and Row, New York.

Konczak, J., Borutta, M. and Dichgans, J. (1997) The development of goal-directed reaching in infants: 2. Learning to produce task-adequate patterns of joint torque. *Brain Res Exp Brain Res* **113**, 465–474.

Konczak, J., Borutta, Topeka, H., Dichgans, J. (1995) The development of goal-directed reaching in infants: hand trajectory formation and joint torque control. *Brain Res Exp Brain Res* **106**, 156–168.

Kuhtz-Buschbeck, J.P., Stolze, H., Jöhnk, K., Boczek-Funcke, A. and Illert, M. (1998) Development of prehension movements in children – a kinematic study. *Brain Res Exp Brain Res* **122**, 424–432.

Lee, D.N. and Aronson, E. (1974) Visual proprioceptive control of standing in human infants. *Percept Psychophys* **15**, 529–532.

Loovis, E.M. and Butterfield, S.A. (2000) Influence of age, sex, and balance on mature skipping by children in grades K-8. *Percept Mot Skills* **90**, 974–978.

Malina R.M. (1980) Biosocial correlates of motor development during infancy and early childhood. In: Greene, L.S. and Johnson, F.E. (eds) *Social and Biological Predictors of Nutritional Status, Physical Growth, and Neurological Development*. Academic Press, New York, pp.50–65.

Marteniuk, R.G., Leavitt, J.L., Mackenzie, C.L. and Athenes, S. (1990) Functional relationship between grasp and transport components in a prehension task. *Hum Mov Sci* **9**, 149–176.

Mason, C.R., Theverapperuma, L.S., Hendrix, C.M. and Ebner, T.J. (2004) Monkey hand postural synergies during reach-to-grasp in the absence of vision of the hand and object. *J Neurophysiol* **91**, 2826–2837.

McFadyen, B., Malouin, F. and Dumas, F. (2001) Anticipatory locomotor control for obstacle avoidance in mid-childhood aged children. *Gait Posture* **13**, 7–16.

McGraw, M.B. (1940) Neuromuscular development of the human infant as exemplified in the achievement of erect locomotion. *J Pediatr* **17**, 747–771.

Payne, V.G. (2001) *Human Motor Development: a lifespan approach*, 5th Edn. McGraw Hill, Boston.

Prechtl, H.F.R. (1986) Prenatal motor development. In: Wade, M.C. and Whiting, H.T.A. (eds) *Motor Development in Children: aspects of coordination and control*. Nighoff, Dordrect, pp.53–64.

Pryde, K.M., Roy, E.A. and Patla, A.E. (1997) Age-related trends in locomotor ability and obstacle avoidance. *Hum Mov Sci* **16**, 507–516.

Ricken, A.X., Savelsbergh, G.J. and Bennett, S.J. (2004) Coordinating degrees of freedom during interceptive actions in children. *Brain Res Exp Brain Res* **156**, 415–421.

Rosenbaum, D.A. (1991) *Human Motor Control*. Academic Press, Inc, San Diego.

Savelsbergh, G.J.P., Davids, K., van der Kamp, J. and Bennett, S. (2003) *Development of Movement Coordination in Children: applications in the field of ergonomics, health sciences and sport*. Taylor and Francis, Oxford.

Savelsbergh, G.J.P., Jongmans, M. and Hopkins, B. (1989) *Reik en Vanggedrag bij Babies*. [Reaching and catching in infants.] Internal report, Faculty of Human Movement Sciences, Free University, Amsterdam.

Savelsbergh, G., von Hofsten, C. and Jonsson, B. (1997) The coupling of head, reach and grasp movement in nine months old infant prehension. *Scand J Psychol* **38**, 325–333.

Schmidtz, C., Martin, N. and Assaiante, C. (2002) Building anticipatory locomotor adjustment during childhood: a kinematic and electromyographic analysis of unloading in children from 4–8 years of age. *Brain Res Exp Brain Res* **142**, 354–364.

Schneiberg, S., Sveistrup, H., McFadyen, B., McKinley, P. and Levin, M.F. (2002) The development of coordination for reach-to-grasp movements in children. *Brain Res Exp Brain Res* **146**, 142–154.

Shirley, M. M. (1931) *The First Two Years: a study of twenty-five babies*. Greenwood Press, Westport, CT.

Shumway-Cook, A. and Woollacott, M. (1985) The growth of stability: postural control from a developmental perspective. *J Mot Behav* **17**, 131–147.

Shumway-Cook, A. and Woollacott, M. (1995) *Motor Control. Theory and practical applications*. Lipincott, Williams and Wilkins, Philadelphia.

Shumway-Cook, A. and Woollacott, M. (2001) *Motor Control: theory and practical applications*. Lippincott Williams and Wilkins, Philadelphia.

Smyth, M.M., Anderson, H.I. and Churchill, A.C. (2001) Visual information and the control of reaching in children: a comparison between children with and without developmental coordination disorder. *J Mot Behav* **33**, 306–320.

Strohmeyer, H.S., Williams, K. and Schaub-George, D. (1991) Developmental sequences for catching a small ball: a pre-longitudinal screening. *Res Q Exerc Sport* **62**, 257–266.

Sveistrup, H. and Woollacott, M. (1996) Longitudinal development of the automatic postural response in infants. *J Mot Behav* **28**, 58–70.

te Velde, V.A., Savelsbergh, G.J., Barela, J.A. and van der Kamp, J. (2003) Safety in road crossing of children with cerebral palsy. *Acta Paediatr* **92**, 1197–1204.

Thelen, E. and Fisher, D.M. (1982) Newborn stepping: an explanation for a 'disappearing reflex'. *Dev Psychol* **18**, 760–775.

Thelen, E. and Fisher, D.M. (1983) From spontaneous to instrumental behavior: Kinematic analysis of movement changes during very early learning. *Child Dev* **54**, 129–140.

Thelen, E., Fisher, D.M. and Ridley-Johnson, R. (1984) The relationship between physical growth and a newborn reflex. *Infant Behav Dev* **7**, 479–493.

Thelen, E., Fisher, D.M., Ridley-Johnson, R. and Griffin, N.J. (1982) Effects of body build and arousal on newborn infant stepping. *Dev Psychobiol* **15**, 447–453.

Thelen, E. and Smith, L.B. (1994) *A Dynamic Systems Approach to the Development of Cognition and Action*. MIT Press, Cambridge, MA.

Tresilian, J. R. and Stelmach, G.E. (1997) Common organization for unimanual and bimanual reach-to-grasp tasks. *Brain Res Exp Brain Res* **115**, 283–299.

Turvey, M.T. (1977) Preliminaries to a theory of action with reference to vision. In: Shaw, R. and Bransford, J. (eds) *Perceiving, Acting, Knowing*. Lawrence Erlbaum, Hillsdale, NJ, pp.211–257.

Twitchell, T.E. (1970) Reflex mechanisms and the development of prehension. In: Connolly, K.J. (ed) *Mechanisms of Motor Skill Development*. Academic Press, London, pp.25–45.

Ulloa, A. and Bullock, D. (2003) A neural network simulating human reach-grasp coordination by continuous updating of vector positioning commands. *Neural Netw* **16**, 1141–1160.

Utley, A. and Astill, S.L. (2007) Developmental sequences of two-handed catching: how do children with and without developmental coordination disorder differ? *Physiother Theory Pract* **23**, 65–82.

Utley A., Steenbergen B. and Sugden, D.A. (2004) The influence of object size on discrete bimanual co-ordination in children with hemiplegic cerebral palsy. *Disabil Rehabil* **26**, 603–613.

van Dellen, T. and Kalverboer, A.F. (1984) Single movement control and information processing, a developmental study. *Behav Brain Res* **12**, 237–238.

van der Kamp, J., Savelsbergh, G.J. and Davis, W.E. (1998) Body-scaled ratio as a control parameter for prehension in 5- to 9-year-old children. *Dev Psychobiol* **33**, 351–361.

van der Meer, A. L. H., van der Weel, F. R. and Lee, D.N. (1995) The functional significance of arm movements in neonates. *Science* **267**, 693–695.

von Hofsten, C. (1980) Predictive reaching for moving objects by human infants. *J Exp Child Psychol* **30**, 369–392.

von Hofsten, C. (1982) Eye hand coordination in newborns. *Dev Psychol* **18**, 450–461.

von Hofsten, C. (1984) Developmental changes in the organization of pre-reaching movements. *Dev Psychol* **20**, 378–388.

von Hofsten, C. (1986) The emergence of manual skills. In: Wade, M.G. and Whiting, H.T.A. (eds) *Motor Development in Children: aspects of coordination and control*. Martinus Nijhoff, Dordrecht, pp.167–187.

von Hofsten, C. (1993) Studying the development of goal-directed behavior. In: Kalverboer, A.F., Hopkins, B. and Geuze, R. (eds) *Motor Development in Early and Later Childhood: longitudinal approaches*. Cambridge University Press, Cambridge, pp. 109–125.

von Hofsten, C. and Lindhagen, K. (1979) Observations on the development of reaching for moving objects. *J Exp Child Psychol* **28**, 158–173.

von Hofsten, C. and Rönnqvist, L. (1988) Preparation for grasping an object: developmental study. *J Exp Psychol Hum Percept* **14**, 610–621.

Wellman, B.L. (1937) Motor achievements of preschool children. *Child Educ* **13**, 311–316.

Whitall, J. (2003) Development of locomotor co-ordination and control in children. In: Savelsbergh, G.J.P., Davids, K., van der Kamp, J. and Bennett, S. (eds) *Development of Movement Coordination in Children: applications in the field of ergonomics, health sciences and sport*. Routledge, New York, pp.251–271.

White, B.L., Castle, P. and Held, R. (1964) Observations on the development of visually directed reaching. *Child Dev* **35**, 349–364.

Wickstrom, R.L. (1977) *Fundamental Motor Patterns*. Lea and Febiger, Philadelphia.

Wickstrom, R.L. (1983) *Fundamental Motor Patterns*, 3rd Edn. Lea and Febiger, Philadelphia.

Williams, J.G. (1992) Catching action: visuomotor adaptations in children. *Percept Mot Skills* **75**, 211–219.

Wing, A.M., Turton, A. and Fraser, C. (1986) Grasp size and accuracy of approach in reaching. *J Mot Behav* **18**, 245–260.

Woollacott, M., Debu, B. and Mowatt, M. (1987) Neuromuscular control of posture in the infant and child: is vision dominant? *J Mot Behav* **19**, 167–186.

Woollacott, M.H. and Sveistrup, H. (1992) Changes in the sequencing and timing of muscle response coordination associated with developmental transitions in balance. *Hum Mov Sci* **11**, 23–36.

Zackowski, K.M., Thach, W.T. Jr and Bastian, A.J. (2002) Cerebellar subjects show impaired coupling of reach and grasp movements. *Brain Res Exp Brain Res* **146**, 511–522.

INDEX